Climate Chaos

Making Art and Politics on a Dying Planet

Neala Schleuning

Minor Compositions 2021

Climate Chaos. Making Art and Politics on a Dying Planet
Neala Schleuning

ISBN 978-1-57027-381-0

Cover image by Alan Montgomery, "Poiesis Disc with Strands"
Cover design by Haduhi Szukis
Interior design by Margaret Killjoy

Released by Minor Compositions 2021
Colchester / New York / Port Watson

Minor Compositions is a series of interventions & provocations drawing from autonomous politics, avant-garde aesthetics, and the revolutions of everyday life.

Minor Compositions is an imprint of Autonomedia
www.minorcompositions.info | minorcompositions@gmail.com

Distributed by Autonomedia
PO Box 568 Williamsburgh Station
Brooklyn, NY 11211

www.autonomedia.org
info@autonomedia.org

"In harsh-edged echo, Titans stir far below. They are all the presences we are not supposed to be seeing – wind gods, hilltop gods, sunset gods – that we train ourselves away from to keep from looking further even though enough of us do."
– Thomas Pynchon, *Gravity's Rainbow*

"This is precisely the time when artists go to work. There is no time for despair, no place for self-pity, no need for silence, no room for fear. We speak, we write, we do language. That is how civilizations heal."
– Toni Morrison

Who the f**k do you think you are? You blame China. You blame India. You blame America. You blame the CEOS, the oil companies, the vague and incoherent "system," the international regulatory regimes, the hypocrisy of the left, the righteousness of the right, the educators, the economy, your parents, your childhood, your job, your bank account, your mental health, your government, everyone and everything but yourself. Wake up! This is no joke. This is absolutely happening and your five planet-lifestyle is the primary cause of it.
– *Adbusters*, Issue 19

CONTENTS

Acknowledgements......................vii
List of Figures........................ ix
Introduction 1
The Changing Sublime 19
Making Art on a Dying Planet............39
Do You See What I Sea?.................73
Nothing Will Ever Be The Same Again 97
Nothing Will Ever Be the Same Again......121
Reworlding 151
Bibliography........................ 161
Endnotes 183

ACKNOWLEDGEMENTS

I AM INDEBTED TO MANY PEOPLE WHO HAVE SUSTAINED ME WITH INSPIRATION and with support when I was doubting my abilities. I am especially grateful way into a future of loving collaboration with the planet.

Special thanks to:

Alan Montgomery for the use of his painting "Poiesis Disc with Strands" on the book's cover.

William Li Campanello for his gift of the book's title;

Art Historian, Dr. Sugata Ray, for opening my heart to the being of water;

The artists who generously granted approval for the use of their images;

Stevphen Shukaitis and publisher Autonomedia for supporting this work and making it accessible worldwide;

Members of a 5 X 5 Research Group on the Anthropocene, sponsored by the Institute for Advanced Study, University of Minnesota; and

Casie Yount for her work on cover design;

For up-close editing and in-depth criticism:

I am indebted to many friends. They challenged and corrected me, suggested new insights, clarified the passages explaining the science of global warming, and some finally just said, "No, no, and no!" when necessary, and found my grammatical errors (well, most of them I hope). They took on this task because they, too, care for the earth, and I am deeply grateful.

Eric Steinmetz, Devorah Molitor, Daniel Edelstyn, Allen Branum, James Koehnline, Hannah Sheridan, Robert I. Yount, Celie Richardson, Fred Whitehead, Kim Blue, Judy LaBrosse, Paul Capel, Matthew Berens, Wayne Nealis.

Individual artists who generously granted permissions

Riber Hansson, Leo Lin, Bill Roberts, Kiakili, Sebastien Thibault, Andrei Popov, James Koehnline, Scott Laserow, Clifford Harper, NomadicAlternatives, Ricardo Levins Morales, Sam Kerson and Katah, Angel Bolígan, Asaf Hanuka

Artists in the Public Domain

Leonardo da Vinci, Katsushika Hokusai, Medieval Map Details, Ivan Bilibin, J.M.W. Turner, Olaf Magnus, Hans Egede, Ivan Aivazovsky, Pierre Denys de Montfort, Utagawa Kuniyoshi

Print Permissions

Timothy Morton, Hyperobjects: Philosophy and Ecology after the End of the World. University of Minnesota Press, 2013.

Excerpt from Walt Whitman, "Song of Myself," from Leaves of Grass. (Public Domain)

Excerpt from David Wagoner's poem, "Lost." In Traveling Light: Collected and New Poems, University of Illinois Press, 1999.

Finally, my deepest thanks to the earth. I listened over a lifetime to the voices of the world around me – the plants and animals, the bees and insects, and the beings inhabiting the deepest oceans. My heart shares in their pain, their losses, and their untimely deaths at our hands. My soul was inspired by their resilience in the face of human destruction and their passion to live and be. I bow in reverence to the gift they give us every day of their presence and their examples of persistence as they move through the seasons of their lives and mine, moving all around me and through me. Maybe this book has made a difference. I hope so.

Every effort has been made to identify copyright holders of visual materials. Errors or omissions have been unavoidable or unintentional. In some cases, I could identify the artist, but was unable to find a contact for him/her after repeated attempts. Images circulating on the internets have been especially difficult to trace and assign appropriate credit, and I apologize for any failures to seek approval.

LIST OF FIGURES

Fig. 1 – Ricardo Levins Morales (United States) – The Planet You're Warming Is Not Yours to Warm

Fig. 2 – James Koehnline (United States) – Burning Midnight Oil Again

Fig. 3 – Asaf Hanuka (Israel) Untitled – Uncle Oil

Fig. 4 – Angel Boligán (Cuba) – Our Flora and Fauna

Fig. 5 – Leonardo da Vinci (Italy) – The Deluge 1517-1518 (Wikimedia commons)

Fig. 6 – J.M.W. Turner (United Kingdom) – Shipwreck of the Minotaur (Public Domain)

Fig. 7 – Ivan Aivazovsky (Russia) – The Ninth Wave (Google Art Project)

Figs. 8 and 9 – Medieval Maps, Detail (2) (Public Domain)

Fig. 10 – Olaf Magnus, 1555 – Sea Serpent (Public Domain)

Fig 11 – Hans Egede, 1734 – Sea Serpent (Public Domain)

Fig. 12 – Katsushika Hokusai (Japan) – Under the Wave off Kanagawa (Kanagawa oki nami ura), also known as The Great Wave, from the series Thirty-six Views of Mount Fuji (Fugaku sanjūrokkei) ca. 1830–32 (The British Museum, Creative Commons

Fig. 13 – Ivan Bilibin (Russia) – The Wave Illustration for Pushkin's Tsar, 1905 (Public Domain)

Fig. 14 – Ricardo Levins Morales (United States) San Ciriaco 1899

Fig. 15 – Pierre Denys de Montfort 1801 – Octopus (Public Domain)

Fig. 16 – Utagawa Kuniyoshi, The umibōzu from the Fifty-Three Parallels for the Tōkaidō, Kuwana Station and the sailor Tokuso, 1843-1845.

Fig. 17 – Andrei Popov (Russia) – Climate Change

Fig. 18 – Riber Hansson (Sweden) – The Ship Earth

Fig. 19 – Leo Lin (Taiwan) – Global Warming

Fig. 20 – Scott Laserow (United States) – Hurricane Katrina 2005

Fig. 21 – Kiakili (United States) – African American Deaths

Fig. 22 – Sam Kerson and Katah (Canada) – Drowning, 2016

Fig. 23 – Bill Roberts Editorial Cartoon Collection, Cleveland State University (United States) Untitled – Tsunami of industrial pollution

Fig. 24 – Sébastien Thibault (United States) –Untitled – Industrial air pollution

Fig. 25 – Clifford Harper (United Kingdom) The Flood

Fig. 26 – Nomadic Alternatives (Canada) Forgotten Tongue

INTRODUCTION

> It may never have occurred to them that so much magic, So much life, might be an indicator of ... awareness. The earth does not speak in words, after all.
> The Earth sees no difference between any of us. ...
> The Earth did not care. ... Humanity is humanity.
> We were all guilty. All complicit in the crime of attempting to enslave the world itself.
> – N. K. Jemisin, *The Stone Sky*

A Cautionary Tale

I BEGIN WITH A SIMPLE STORY, A CONCEIT BASED ON AN EMBELLISHMENT OF two interrelated Greek myths of the gods – Zeus, (the Big Guy), his colleague, Prometheus, and Prometheus' daughter, Pandora. Prometheus was a Titan who loved humankind. He stole fire from the heavens and brought it to humans in defiance of Zeus. He was punished horribly for his defiance, sentenced to live out eternity having his liver eaten out daily by an eagle. We don't often connect the myth of Pandora to Prometheus, but it is important, because she was his daughter, and Zeus was still angry. Zeus further punished Prometheus by giving Pandora (the first woman) to his brother. One day, while her husband was away, Pandora found a jar (the "box") and opened it. Unimagined horrors were released from the jar, scattering across the world. As she quickly closed the jar, she noticed that only Hope remained. We are usually reassured and uplifted by the term hope, but according to translators, the Greek word for the element remaining in the jar could also "have the pessimistic meaning of 'deceptive expectation.'"[1] A deceptive expectation could mean, of course, that hope might be delusional.

The two tales together weave a cautionary tale of the rise and fall of humanity. Prometheus' gift of fire is often interpreted as the gift more generally of technology. We were given the gift of technological "making," of manipulating the environment (power over nature). Subsequently, as we are coming to learn, our making turned out to be our undoing, which the tale of Pandora exemplifies. We are, it seems, often the creators of our own horrors. We have discovered that we are the Anthropocene.

And so begins my story of the Anthropocene, how our altering the natural world has become our undoing. This book is a call for urgent action on this existential threat facing the human race. This book also brings to you another warning, another apocalyptic tale. The narrative begins with a dire warning of a self-inflicted climate crisis. How it ends depends upon how humanity will collectively shape a new relationship with the planet.

Just What Is This "Anthropocene?"

The scientific term Anthropocene combined two Greek words, "anthropos" (human) and "cene" (recent) and was first formally coined in 2000 as a geologic epoch describing humanity's impact on the environment of the entire planet. While previous "cenes" like the Pleistocene and the Holocene in geologic time frames were characterized by the various animal and plant life indigenous to those epochs, the Anthropocene is different: it focuses on the significant impact on geologic processes by one animal on the planet – the human animal. We have become the principal agent of change on the surface of the planet. In *The Climate of History: Four Theses,* historian and physicist Dipesh Chakrabarty called for a recognition of the uniquely permanent and ongoing human impact on the environment:

> To call human beings geological agents is to scale up our imagination of the human. Humans are biological agents, both collectively and as individuals. They have always been so. There was no point in human history when humans were not biological agents. But we can become geological agents only historically and collectively, that is, when we have reached numbers and invented technologies that are on a scale large enough to have an impact on the planet itself. To call ourselves geological agents is to attribute to us a force on the same scale as that released at other times when there has been a mass extinction of species.[2]

An early reference to the far-ranging impact of human activity on the planet can be traced to Russian geologist Aleskei Pavlov's use of the term "anthropogenic" in 1922, according to *The Great Russian Encyclopedia*. This concept was similar to the term Anthropocene, which was popularized by Nobel Prize winning atmospheric chemist Paul J. Crutzen in 2000 to describe the magnitude of the impact of humans on the planet. This new term was selected because the geologic record is our primary means of cataloging the measurable changes generated by natural processes, and more recently, by the dramatic changes initiated by human activity. While humankind has always altered the environment, the recent changes called for a re-definition that reflected the existential threat to all life.

Traditionally, to establish an official geologic time period (in this case, an epoch), scientists have determined that there must be some geologic marker that can permanently identify the period. Scientists still have not agreed upon exactly when the Anthropocene began, but there is no doubt that human activities have permanently impacted the planet, and in a way that threatens the continued existence of the human race itself. Did the Anthropocene begin with the migration of seeds and changes to the biome with the advent of agriculture around the globe? Did it begin in the nineteenth century with the industrial revolution and the traceability of fossil fuel pollution? Or does it reflect a more recent time period, the atomic age and the fallout of its bright isotope markers of nuclear fission and fusion events that traveled around the globe?

We are now in the midst of the Anthropocene, and its effects are still intensifying. These changes have all happened quickly. Whenever it began, changes that normally might have taken place over tens or hundreds of thousands of years have rapidly accelerated in recent centuries. The last half of the twentieth century saw some of the most rapidly escalating changes. This time period is referred to in the literature as "The Great Acceleration" – the period after the end of the Second World War when economic activity expanded exponentially, when the consumer culture emerged, driven by the unleashing of our human desire to accumulate the latest and newest product, and when human populations skyrocketed with breakthroughs in medicine and general hygiene. The Acceleration has brought us to point where we must put our foot on the brake.

Another important event marking the parameter of the Anthropocene is the ongoing and accelerating extinction of species begun in the Holocene. This activity is referred to as the Sixth Mass Extinction. As human activity was expanded everywhere on the planet, the other beings occupying various ecological niches have been driven out and exterminated as they clash with humankind's needs and wants for resources and space.

The Anthropocene is not just some abstract scientific theory to think about. It is an ongoing and accelerating phenomenon that is being carefully measured and monitored by scientists around the world. Their key measurement marker is registered by changes in the temperature of the planet as a whole. As the temperature rises inexorably, a point will eventually be reached where human survival is threatened. We cannot live on a hot planet. Regular reports on carbon dioxide (CO2) levels in the atmosphere are released by the United Nations Intergovernmental Panel on Climate Change (IPCC) that was created in 1988. In 2018 the IPCC issued a special report, setting a deadline of 2030 as the year by which we need to reduce carbon emissions by 45 percent, in order to keep on track with longer term efforts to slow the temperature rise.

Since the onset of the Industrial Revolution, the Earth has undergone an unprecedented rise in the global average temperature. Despite continuous variations in Earth's climate over the past millennia, the rate and extent of the current warming trend has caused considerable concern within the scientific community.[3] Over ninety-seven percent of climate scientists agree that global warming over the last century has been caused by human activities, primarily through the release of greenhouse gases to the atmosphere. Unless we take immediate action, such activities will put the Earth on a catastrophic trajectory to destroy life as we know it.[4] The Paris Agreement (2015) – a major, formal and global agreement – addressed the impact of human fossil fuel use on the environment. It is perhaps the closest the world has ever come to making a significant international agreement for the environment. The Paris Agreement was ratified by 185 of 197 countries of the world and created the framework for the first concerted global effort to reduce levels of carbon dioxide, a by-product of fossil fuel consumption and a greenhouse gas, in the atmosphere. Other countries are in the process of ratification, with the exception of the United States which has signaled its intent to withdraw effective November 4, 2020.[5] The Paris Agreement, however, is nonbinding and each country is free to establish its own carbon dioxide reduction targets.

Fossil fuel use is raising the temperature of the planet by pumping more carbon dioxide (CO2) into the atmosphere. Data of global CO2 averages across land and water provide a stark overview of the current Earth temperature and the projected impact of increases in the concentration of atmospheric CO2. A pre-industrial baseline from which to measure changes has not yet been officially determined, but studies of ice core samples indicate that during the last ice age (c. 100000–10000 years ago), CO2 concentrations measured approximately 200 ppm (parts per million). Pre-industrial levels have been estimated at 278 ppm. Since 1979, global data has been systematically collected. In January 1979, measurements of 336 ppm were found. The Mauna Loa volcanic eruption in 1989 increased CO2 levels to 350 ppm, and Mauna Loa remains the site with the longest record of CO2 levels. The amount of CO2 in the atmosphere continues to climb and by December 2019 reached 412 ppm.[6]

While it is extremely difficult to estimate temperature changes during past geologic eras, it is known that since the nineteenth century temperatures have continued to rise. This rise is likely due to industrialization. When the concentration of atmospheric CO2 increases, the temperatures also increase; a phenomenon known as the greenhouse effect. According to the National Aeronautics and Space Administration (NASA), the average global temperature during the twentieth century was 53.6 °F, and "[s]cientists have high confidence that global temperatures will continue to rise for decades to come, largely due to greenhouse gases produced by human activities. The Intergovernmental Panel on Climate Change (IPCC), which includes more than 1,300 scientists from the United States and other countries, forecasts a temperature rise of 2.5 to 10 degrees Fahrenheit over the next century."[7] The Paris Agreement has set a goal of keeping the increase in temperature below 3.6°F over that time period.

The first step of the Agreement is to limit the global temperature increase to 2.7 °F by reducing carbon emissions. Nevertheless, that limit is quickly being approached. In a 2018 prediction by the United Nations, it was estimated that this limit would be reached as early as 2040.[8] These increases may seem small (what's three degrees, why is that an issue?), but the Earth has maintained relatively consistent levels of atmospheric CO2 (100–200 ppm) for millennia. Thus, the Earth's inhabitants have become accustomed to pre-industrial conditions, such that even small changes in temperature will upset the entire system, possibly leading to mass extinctions. "The Intergovernmental Panel on Climate Change [IPCC] estimates that 20 to 30 percent of assessed plants and animals could be at risk of extinction if average global temperatures reach the projected levels by 2100. Evolution would have to occur 10,000 times faster than it typically does in order for most species to adapt and avoid extinction."[9] According to David Biello, "A new analysis of the temperature and fossil records over the past 5 million years reveals that the end of the Permian is not alone in this association: global warming is consistently associated with planetwide die-offs."[10]

The temperatures alone do not tell the whole story – it is also the rate of increase that is important. The National Oceanic and Atmospheric Administration (NOAA) has reported that between 2011 and 2012, CO2 emissions from fossil fuels (e.g., coal, oil, and natural gas) increased 9-10 ppm and the rate of increase is predicted to accelerate.[11] Because increases of CO2 emissions cause tandem increases in temperatures, the goal will be to limit our emissions. The longer we delay taking action, the more difficult, and expensive, it will become to mitigate the detrimental effects caused by human activities. Currently, we have already begun to witness the resulting harmful effects on the environment such as an increase in extreme weather events, ocean acidification, decreased food production, and loss of biodiversity. Without mitigation it has been predicted that by 2050, concentrations of CO2 will be 550 ppm.[12]

One of the most contentious arguments in the global debate over how to proceed is the question of who is responsible for carbon dioxide pollution that caused this increase in temperature. Further, the nations that historically created the Anthropocene are being challenged with the argument that they have an obligation to financially accommodate developing country's desires to grow their more recently industrializing economies. The first countries to industrialize have been the largest contributors to the current dire situation.

> The United States, Canada, Europe (and Eurasia), Japan, and Australia have together contributed around 61 percent of the total [CO2 emissions], as compared to 13 percent for China and India taken together. Russia accounts for another 7 percent, and world ship and air transport are 4 percent. The entire rest of the globe accounts for 15 percent of cumulative emissions.[13]

The nations that historically created the Anthropocene are being challenged with the argument that they have an obligation to financially accommodate developing countries' desire to grow their more recently industrializing economies. With the need to set limits to continued industrial expansion and consumption by all economies, the stage is set for tense debates about who should make the steepest cuts in order to allow others to achieve the same level of development. This was a major source of contention in the 2019 Climate Change Conference (CO25). The data are clear, as you can see in this CarbonBrief YouTube video released in 2019.[14]

Scientists have long known that human activity is raising the temperature of the planet. Some of those early scientists and philosophers contributed ideas that frame our understanding of the intimate relationship between humans and nature. The term "biosphere" was first popularized by Russian scientist Vladimir Vernadsky. The biosphere, as Vernadsky explained, is the thin surface of the earth where all life forms exist and interact. It describes the complex interactive systems that sustain all life. The term has since entered the environmental lexicon to describe the unitary nature of planetary life, and the synergy between the natural world and human activity.

Vernadsky also made a profound observation about the force of human beings in shaping the environment, crediting Russian geologist Aleksei Petrovich Pavlov and others for their insights: "A.P. Pavlov (1854-1929) in the last years of his life used to speak of the *anthropogenic era* in which we now live. ... [H]e rightfully emphasized that man, under our very eyes, is becoming a mighty and ever-growing geological force." Vernadsky proposed calling the next planetary evolutionary step the noösphere – the third phase of life on earth, following inanimate life and the biosphere (noos = the Greek word for mind).

The noösphere reflected the supremacy of the human mind interacting with all life.[15]

Vernadsky's framing of the biosphere anticipated the resonant formulation of English scientist James Lovelock's Gaia hypothesis – that inanimate and animate life forms on the planet are one single life organism. Lovelock continues to be a prominent voice in the environmental movement, and his ideas permeate all of our thinking about nature and how to move forward to remediate our damages to the environment. This global consciousness contributed to the contemporary framing of the Anthropocene.

In the 1970s, a new term, Deep Ecology, reflected the shift in thinking to a more ecocentric (earth-centered) philosophy, more sharply focusing theory and debate on the needs of nature. The expansion in thinking was inspired by the writings of Henry David Thoreau, Lev Tolstoy, John Muir, James Lovelock and especially Norwegian philosopher Arne Naess. Based on these ideas, the modern environmental movement gradually moved the debate about the relationship between humankind and the earth to center stage.

How This Study Began

How I came to care so deeply for the planet is important to share at this point, if, perhaps, only to inspire people who are just learning to appreciate the world's effects on their own individual lives and communities. Each of us, in some way I believe, has to develop an intimate and personal connection with the richness of life around us.

I was always interested in nature, even as a child. I was a Girl Scout, and over years accumulated a lot of "badges" recognizing my studies about the natural world. I studied the classification of rocks and bugs and plants and animals, and even the stars in the sky. Over subsequent decades I planted gardens, collected samples of rocks, gathered morel mushrooms in the woods, and camped in the deep woods of northern Minnesota.

I lived in the southern Minnesota countryside for many years, walking my dog along the gravel road, growing a garden, listening to the cottonwood trees talk to me and watching the birds build their nests in the underbrush. Wild turkeys, deer, woodchucks, raccoons, and the many birds shared my world. At one point my walks got boring, so I took up the study of grasses growing in the ditches alongside the road to better understand the world that surrounded me. I identified many plants that turned out to be the "old" grasses – native plants that had been cleared from their natural habitat and replaced with fields of corn and soybeans. The native grass seeds survived by hanging along the roadsides. I learned that there are over 10,000 grasses on the planet – my first deep awareness of global life. The scope and diversity of the grasses, and indeed all life, is staggering.

In my twenties, I first became aware of environmental damages caused by human activity and how the affected my own life. In the 1940s and 1950s, nuclear weapons testing and the catastrophes of Nagasaki and Hiroshima brought home the horrors of worldwide radiation poisoning. In 1962, Rachel Carson's book *Silent Spring* alarmed the entire world to the dangers of chemical pollution. Her warning resulted in the banning of DDT. Since then, the list of herbicides, pesticides, and pharmaceuticals polluting our environment has grown exponentially. Later, throughout the Vietnam War, we saw the devastation wreaked on that country's ecology and the human gene pool by Monsanto's Agent Orange and similar defoliant chemicals produced by Dow Chemical and other companies. Veterans returning from Vietnam were often the first indication of the widespread toxic effects on human populations and the environment. The recent use of depleted uranium weaponry in the Middle East also planted the devastating seeds of on-going genetic mutations in civilian victims. Our collective genetic history was changed, and continues to change in response to the poisons we have created in our world.

In 1984, the world's largest chemical disaster to date awakened the whole world to the dangers of chemical production. Infrastructure failure at Union Carbide India Limited gas plant in Bhopal, India exposed over half a million people to the gas methyl isocyanate (MIC) – resulting in thousands of deaths and hundreds of thousands of injured. The fallout continues to this day, including ongoing health problems for those exposed and environmental contamination of ground water and soil. The disaster subsequently energized international activism.

Learning and studying further opened my eyes. In response to the cascading environmental crises, we saw increases and improvements in environmental education and collective action. Earth Day was established in 1970, replacing the earth holiday of my youth, Arbor Day. The shift in perspective from planting trees to caring for the entire planet was a profound transition in thinking for the whole society. In the early 1970s, I took my first course at the University of Minnesota that specifically focused on ecology and I directed my attention to the economics and politics of our relationship with nature. Minnesota writer Meridel Le Sueur's *oeuvre* grounded my deep passion for nature in my home place of Minnesota and the Midwest – the tallgrass prairie.

My interest in the environment led to engaging with the politics of the environment. I soon came to learn that all environment issues are local issues first. I'm a tree-hugger, I confess, and over the years the defense of the earth has become a priority. As the devastation has increased, so too, has my commitment to seeking alternatives. In the 1990s I worked with my neighbors to keep the Minnesota National Guard from taking over a small state park and turning it into a shooting range. The southern Minnesota county I lived in had fought a long battle decades before to save a local native prairie. Another local organization, Save the Kasota Prairie, inspired our actions. My neighbors and I won this small battle, and today, American bison roam throughout the park. In 2017, mass protests over a proposed oil pipeline across the Standing Rock Reservation in South Dakota, awakened the whole world to the life and death struggle between the fossil fuel culture and the culture of life. "Mini Wiconi" – the Dakota words for Water Is Life – became the rallying cry for people across the globe committed to saving the planet and the precious resources that sustain human life.

These are just a few examples of my personal journey of educating myself and acting on behalf of the environment. My small actions were joined with the millions of people worldwide struggling to save their trees, their salmon streams, their towns, their neighborhoods, their drinking water, their farmland – all grounded in the real world of living nature.

Environmental Awareness Intensifies

In response to the growing environmental awareness of pollution, in the last third of the twentieth century, there was hope that the world would take up the task of repairing damages and protecting the environment. In the United States, a spate of legislative actions were rapidly undertaken at the federal level. The Clean Air Act was passed in 1963 to attack air pollution. The United States established the Environmental Protection Agency in 1970. Rachel Carson's call to ban DDT was finally enacted and the Clean Water Act followed in 1972. The Endangered Species Act was passed in 1973. Many states and municipalities followed suit, seeking to protect and/or remediate local resources by setting up pollution control agencies to enforce the new legislation or establish even stricter measures, such as those of California and several other states that require stricter controls of auto emissions. And above ground nuclear testing was banned worldwide; offshore drilling for oil was curtailed in the United States east and west coasts.

As concern about the environment grew, so did government regulation of pollution. Environmental attorneys have been in court for the past thirty or forty years protecting the environment. In recent decades, the nation has been gradually transitioning to alternative energy sources. In Minnesota, for example, the legislature required the local energy company to annually transition a percentage of electrical power production to wind or solar. However, with the election of Donald Trump as president of the United States in 2016, many of these efforts to curb and eliminate pollution have been reversed. The *New York Times* reported in the fall of 2019, that sixty-seven environment regulations had been cancelled. The systematic removal of environmental regulations preserving the natural world and all species was met by heightened political actions globally on behalf of the environment. The withdrawal of the United

States from the Paris Agreement raised the ante even higher for the global community.

Over time, there has been a growing conviction globally that the political challenges of the Anthropocene had to extend far beyond tree-sits and monkey-wrenching to save primeval forests; beyond protests against Nestlé drawing down vital water resources for private profit; and beyond recycling (although all of these acts deserve our praise and merit our involvement). Rising sea levels, California and Brazil on fire, and frequent "thousand-year" storms and hurricanes, have upped the ante further. There was something bigger happening, greater forces overwhelming personal efforts. Gradually I came to see that the problems were systemic. And actually quite simple to resolve – although we have not yet made that commitment. We have to drastically reduce the burning of fossil fuels which means switching from driving our cars and trucks to adopting more public transportation options; we must find an alternative economic model to capitalism and its commitment to continually expanding growth and exploitation of the natural environment; and if we do not abandon our personal economies of accumulation and consumption, the human race will end, to borrow a phrase from T. S. Eliot, "not with a bang, but a whimper."

How Does Art Fit Into This Narrative?

My interest in linking art and the challenges of the Anthropocene was first ignited in a course offered by the Art History Department at the University of Minnesota. Dr. Sugata Ray, on loan for a year to the University of Minnesota from Berkeley, was the instructor of the course "Liquescent Materiality: The Cultural History of Water." It focused on how various cultures conceptualize and represent water and its power as a force of nature, its agency. The course was timely and appropriate in its insights into global warming and the Anthropocene. As we studied the role and place of water in art and culture, the challenge for me was how to apply that knowledge to our current concerns over all aspects of the environment. How do we imagine nature have agency, acting with intent? What images might convey the huge concept of "global warming?" What stories should we tell one another about the pollution in our neighborhoods? How can we inspire hope, as dread takes center stage with every disaster?

I was particularly fascinated with the readings and discussion around the agency of the sea and how human societies across time have developed myths and legends about the sea and its power. The sea became, in my mind, a useful metaphor for all of nature. You will see it appear again and again throughout this book, as a representation of the power of nature generally. As I began collecting and exploring political images of the sea and the environment generally, several themes gradually emerged that shaped my thinking about the agency of nature in general, and of the sea specifically.

My interest in the relationship between art and politics is captured in my 2013 book, *Artpolitik: Social Anarchist Aesthetics in an Age of Fragmentation*. It was an exploration of the power of visual representation to communicate complex political ideas that could motivate people to action. As the twentieth century progressed, the world was awash in visual imagery, and the images took on an increasingly significant role in shaping the reality in which we all live. In such an environment, it becomes even more important to understand the nature of imagery, of its impact on our lives, and of how we might play a greater role in controlling and countering the images that surround us. It is in the world of visual representation that the battle for the hearts and minds of people must be fought and won.

Similarly, politics is a mechanism for sharing symbols and affirming collective meaning. Art and politics are two of the most powerful mechanisms for visioning, for inventing our future. Together, they hold the key to determining what kind of world we will create. Communicating the threats to the planet and to humanity via artistic expression – essential to our understanding of the Anthropocene – seemed to be the obvious next step in my thinking.

A Narrative for the Future

A new narrative, with new words and new visions is emerging – a narrative based on a dialogic relationship with the earth and its other beings. We don't need to idealize the other beings, we need to live alongside them in a negotiated arrangement. As will be argued throughout this book by many writers and thinkers, the idea of Nature (with a capital N) is now dead. This was an Enlightenment notion – that "we" (the human subject) are on one side of an equation, and that "they" and "it" (nature) are on the other, with one important difference: human beings are in charge.

There is reason for optimism. Over the past fifty years, even as the devastation of the natural world continues unabated, we have already begun to shape this new consciousness, to develop a new narrative. The seeds for this new vision were planted in the early environment movements, in the need to respond to the emerging threats of urban smog, the Santa Barbara oil spill, and the burning of the Cuyahoga River in Ohio in 1969. Public awareness made clear that humankind was threatened by the work of our own hands and accelerating economic development. As we confronted these challenges to the quality of our air, our water ways, and their impact on our personal health, the contributions of scientists worldwide illuminated our interconnected world. The new reality is a story of enmeshment, of relationship. We are nature, nature is us; there is no separation, only the effort to make further connections.

We have a collective obligation to address the reality of humankind's role in creating the Anthropocene. At minimum, we owe reparations for the damages done to our children, and the generations beyond our children, and to the other beings with whom we share this planetary home. Going forward, this book asks the question: What is the story of the Anthropocene and how do we best tell it through visual art? How do we engage others to take up the task of changing the world for the better? The answers will revolve around the environmental stories we tell one another. In an interview with writer Bill Kilby, psychologist Per Espen Stoknes, outlined the kinds of stories we need to begin sharing with one another: "a) green growth opportunities, b) better quality of life, i.e. what does a low-carbon society look like? c) the ethical stewardship story, and finally, d) stories on re-wilding and the resilience of nature. The more people start believing we can create a better society with lower emissions, the sooner they can start taking action."[16]

The telling of a tale is central to conveying complex ideas that people will embrace and act upon. The central role of storytelling emerged in the deep time of human history, before the written word. Over time, the sharing of narratives shifted from oral transmission to the written word. Visual art, too, adopted the narrative form, either using representation to solidify familiar narratives such as icons of the Virgin Mary, monuments to war heroes, or designing propaganda to shape collective thinking.

The setting for many stories was often the natural world. As philosopher David Abram emphasized in *The Spell of the Sensuous*, place is always important to story: "[T]he sensuous world – the world of our direct, unmediated interactions – is always local."[17] Arguing that the intimacy of a relationship with the natural world has been severed by the introduction of written language, Abram traced the transition across many cultures from intimate, personal connections with the natural world, through symbolic representations, to oral sharing of language, to the creation of written alphabets that are a total abstraction away from the felt, and experienced, real. This book challenges us to return to the reality of the natural world – to reground ourselves and our societies in the true source of our livelihoods.

We have, of course, always created narratives about nature. Early in the twentieth century, the United States avidly embraced the idea of a "wilderness" – the belief that pockets of wild lands should be cordoned off to retain their original character. During the presidency of Theodore "Teddy" Roosevelt, the United States took its first steps in preserving wilderness with the establishment of the National Park System. A romantic narrative of the environment lauded the wonder and beauty of wild places, untouched by humankind. It was as if we somehow believed we could build a dome over the national parks, protecting them from radioactive fallout, air and water pollution, or a rapidly expanding human population. The idea of wilderness, however, did not reflect the complex reality of the nature/human relationship. While parts of the natural world were preserved inviolate, the balance was open for "development" and exploitation. Unfortunately, in recent years, even those minimal protections have been weakened. Federal government policies are increasingly encouraging the exploitation of national resources on public lands where mining, grazing, clear-cutting of state and national forests, and fossil fuel extraction are escalating.

In the same manner, to this day Americans continue to idealize certain species of animals as we systematically drive them to extinction: the wolves, elephants, tigers and lions, the polar bears, grizzly bears, and even the gentle black bears. It seems that the more we destroy these apex predators, the more we idealize them, preserving a few for our pleasure in zoos and sanctuaries. We seem unable to acknowledge that we are not really interested in "saving" the polar bears, because to save them, we would need to save their Arctic environment, which would require a global revolution. It seems to me, too, that other than the polar bear's Arctic home, and the orangutan's rainforest home, we subconsciously ignore the environments needed for sustainability by all other endangered species, including ourselves. We know the ice is melting because our activities are melting it and we refuse to reduce our fossil fuel consumption; we can't really save elephants because capitalism coupled with misguided beliefs in medical magics guarantees their extinction; we know our french fries are prepared in palm oil grown on plantations destroying the habitat of orangutans; and Americans are quite comfortable dumping runoff from herbicides and pesticides into our lakes and rivers because we like the look of a weed-free lawn. And the bees. We all know what we're doing to the bees, don't we?

We have, it seems, avidly avoided seeing our complicity with the destruction unfolding around us. In a capitalist world there is no "saving" for the future, no "husbanding" of resources, and no restraining what Thorstein Veblen called our "conspicuous consumption."[18] There is only ongoing exploitation until nothing is left.

This all has to change.

Outline of This Book

Several key ideas will be explored in the following pages: the relationships between the philosophy of aesthetics and artistic practice, between art and politics, and between several political pathways forward. The chapters exploring these fields are of equal significance, and form the core of my argument. Three chapters are focused on art, and two explore politics.

Chapter One outlines the history of the idea of the sublime. It is the foundation for subsequent arguments I will make about the relationship between art and the natural world. It is probably the densest chapter in terms of content, and you might find it difficult going. I know I did, writing it! I can only encourage you to follow the thread of the philosophical argument because how we think about things affects how we eventually understand, and then act upon our thoughts. Please struggle through this chapter with me.

The idea of the sublime has its roots in the first century, but my focus is on its expression as the dominant aesthetic concept in western art since the eighteenth century. The sublime is not just a way to describe an experience of beauty, although we can see the sublime expressed in the art of Romanticism, where it conveys the idea of grandeur, beauty and pleasure beyond comprehension. But there is also a dark side of the sublime, an undercurrent that reaches to the heart of contemporary environmental aesthetic experience. The ancient roots of the sublime lie in humanity's attempt to articulate a visceral response to the primitive fear and respect for the natural world. The aesthetic and ethical arguments by five key philosophers are presented to make my case for a new way of thinking about art and the sublime experience in the context of the natural world: Edmund Burke, Emmanuel Kant, Theodor Adorno, Timothy Morton, and Emmanuel Levinas. Understanding the sublime will help us develop an aesthetic response to the Anthropocene.

Chapter Two explores artistic practice, examining perspectives that artists should consider as they seek to articulate a new sublime to address the Anthropocene. Several themes will stress the importance of creating a narrative; the need to frame the difficult but necessary awareness of the existential threat of the Anthropocene and how our self-preservation is at risk if we do not look deeply into our cultural avoidance of fear and death; and especially a new intellectual framework that rethinks our place as human beings in relation to the other beings with whom we share this planetary home.

In Chapter Three, I will share examples of contemporary art of the Anthropocene that are widely shared on social media where images of the sea are often used to educate and alert us to the need for new thinking and action about the environment.

Various political responses to the Anthropocene will be explored in Chapters Four and Five. All of the scientific knowledge, political insights, and artistic talents focusing on the Anthropocene will have to be turned into sustained action – now and long into the future. How we face the hard reality that things will never be the same again as the current and unsustainable economic structures crumble will be humankind's greatest political challenge. How politics will guide those responses will be crucial for creating an environmentally sustainable future. We will need to develop new skills and new ways of living together.

A major thread tying the entire book together is the imperative to convey a sense of urgency. I am afraid for the future, and I believe you should be, too. Our actions today will mean life for our descendants, for the polar bears and the wolves and the tigers; for the bees; for the fish in the ocean; for all of us – even for the air and the water. I am one small voice, and we need to engage the voices and actions of billions. I also feel a personal and generational obligation to take up the challenge – my generation was responsible for the Great Acceleration – the period since the end of the Second World War during which the population expanded, consumerism exploded, and insatiable consumption of fossil fuels began. In a recent communication, cultural theorist and filmmaker Michael Truscello concluded that

> The Sixties Generation will be known as the greatest failure in the history of the Left, eventually. They had the resources and the numbers, and they existed in a pre-Internet context. No generation will ever again be as materially rich as the hippie generation (in part, I base this on the science of the Great Acceleration). All generations from this point on will be trying to create revolution at a serious deficit of people and resources, and we must deal with the surveillance and counter-insurgency apparatus known as the Internet. And of course contemporary revolutions occur within a mass extinction event.[19]

When I speak to friends of my generation, I often hear, "Well, I don't have to worry, I'll be dead." That's not good enough. None of us can just walk away from the problems of our shortsighted thinking and acting. So this book is my call to arms to my generation and all generations into the future. My message on the wall: "Something Must be Done." It goes without saying that subsequent generations must also become engaged to take up this long battle for our survival.

This book turned out to be a lot darker than I had anticipated. As the news about the damages the earth had sustained continues unabated, and our collective will to begin efforts to remediate the damage is painfully slow to materialize, I grow increasingly alarmed. I had originally hoped – as likely we all have been hoping – that there would be a fairly seamless transition from one well-thought out plan to another. World governments would put some real teeth into the Paris Agreement; alternative energy sources would rapidly replace fossil fuels; and science and technology would save us somehow. But solutions are not forthcoming as rapidly as they are needed. The largest economy on the planet has withdrawn from the Paris Agreement; tariff wars are raising the cost of solar panels; oil extraction has been ramped up with cascading environmental damages resulting from the fracking and tar sands industries; and the bees are still dying.

As we fiddle while "Rome" and the vast forests of the Taiga, the Amazon, Africa and the entire continent of Australia are burning, climate scientists tell us that there is far less time to make an easy transition as the CO2 levels continue to rise dramatically. We are already seeing cascading effects – rising sea levels, human migrations from drought areas resulting in increasing tribalism and fear and hatred of the "other," declining nutrition in crops as temperatures rise.

We have to get busy, and we have to get busy quickly. Change will happen whether we are ready or not. But we still have a little time to make some of the "right" decisions and initiate forward movement. If we don't, the catastrophe(s) will continue to unfold.

A Final Note: "Global Warming" or "Climate Change?"

The words we use to tell our stories for the future are important to conveying the message of the Anthropocene. George Monbiot recently reflected on the power of our words and how we use them to shape the story of our relationship with the environment: "Our assaults on life and beauty are also sanitised and disguised by the words we use. When a species is obliterated by people, we use the term 'extinction,'" which seems to lay the blame for disappearance on the species rather than just say we killed them, or murdered, or destroyed them. He also examined the terminology describing the climate crisis itself, preferring "global warming" over the term "climate change." "'Climate change' also confuses natural variation with the catastrophic disruption we cause: a confusion deliberately exploited by those who deny our role. (Even the neutral term 'climate change' has now been banned from use in the US Department of Agriculture.)"[20]

Physicist and climate expert Joe Romm concurred:

> In new polling by the Climate Change Communication efforts of Yale and George Mason, 'global warming' is the winner – across the board. We found that the term *global warming* is associated with greater public understanding, emotional engagement, and support for personal and national action than the term *climate change*. ... the use of the term *climate change* appears to actually *reduce* issue engagement by Democrats, Independents, liberals, and moderates, as well as a variety of subgroups within American society, including men, women, minorities, different generations, and across political and partisan lines.[21]

Environmental philosopher Timothy Morton, too, prefers the term global warming. "What we desperately need in an appropriate level of shock and anxiety concerning a specific ecological trauma – indeed, the ecological trauma of our age, the very thing that defines the Anthropocene as such."[22] And in 2019, the British newspaper *The Guardian,* changed the newspaper's style sheet to reflect even more accuracy: the paper now refers to climate change as the "climate crisis," and global warming is called "global heating" in all their reporting.

Fig. 1: Ricardo Levins Morales

THE CHANGING SUBLIME
EMBRACING THE ANTHROPOCENE

> All human beings are descendants of tribal people who were spiritually alive, intimately in love with the natural world, children of Mother Earth. When we were tribal people, we knew who we were, we knew where we were, and we knew our purpose. This sacred perception of reality remains alive and well in our genetic memory. We carry it inside of us, usually in a dusty box in the mind's attic, but it is accessible.
> – John Trudell

What Is the Sublime?

THINKING ABOUT ART IN THE ANTHROPOCENE BEGINS WITH UNDERSTANDING how we think and feel about the environment, and about how we convey those thoughts and feelings in artistic expression. This chapter will begin with an examination of the changing aesthetic concept of the sublime and its application to the practice of art. Most of us are somewhat familiar with the term sublime. For example, we might say to a friend, "That dessert you made last night was absolutely sublime!" Or, while on vacation, we might say: "The view from the

North Rim of the Grand Canyon was sublime!" The sublime is commonly interpreted as a pleasurable experience of grandeur, great beauty, or awe-inspiring expansiveness. And there is also a quality of uncertainty at the heart of the sublime, especially when it is difficult to find the right words to articulate our responses to what is happening. We are amazed, struck dumb.

My early personal experiences of the sublime were linked to water and landscape, lakes and the vast prairies of the Midwest. I stood in awe and wonder on the shores of Lake Michigan and the vast deeps of Lake Superior, enthralled with the incessant movement of water and light. In a poem I wrote many years ago, the prairie was my seascape, where my vision could expand:

> I love this prairie. The earth's soul spread wide and deep, the ecstasy of sky and space, horizon and embrace.
>
> I love this place
>   Thin line between earth and sky.
>   Between here and there, far and near, love and hope.
>   Past and future all directions.
>
> A fragile grace, reforming as it moves. This line between, threatening at each movement to break up. I lose myself again, vertigo, balance breathless on the wind.
>
> A sea. A sea of dreams and sky.

Coming face to face with the sea off the coast of northern Florida was an entirely different experience:

> I remember the first time I saw the sea. The vast expanse, the almost there, almost falling off the edge of the world.
>
> We grip the edge, balanced on this abyss of space.
>
> Some deep instinctive fear. The deep knowledge of no limits. The moving shoreline, end of land and the beginning of what? Perhaps this is prairie. Perhaps this is love. Perhaps this is freedom.
>
> This is the real space, this space between, the space of shaping and re-forming, of doing and undoing, of being and becoming, of stretching, reaching, making. The place where the stars dip and turn, their vast array in full bloom.

Balanced on the edge of sky, I cling to the earth. The eagle
watches.[1]

Through the words of the poem, I moved from observer to participant, from the safe distance of awe and amazement to teetering on the edge of disorientation, uncertainty and fear. As I stood there on the wet sand, enjoying the waves rolling in and admiring the immensity before me, the waves pulled the sand from beneath my feet – a moment of instability bordered on panic, and a great fear of the sea overwhelmed me. The sea becomes a foreboding presence. This is the sublime that is most familiar to us, and it dates back to the seventeenth and eighteenth century, the era we call the Enlightenment.

The idea of the sublime has shifted over time, dating back to the writings attributed to a Greek philosopher, Longinus, in the first century. Over the centuries, definitions of the sublime have always been articulated in the context of the relationship between humankind and the natural world. Explorations of this concept accelerated in the eighteenth century when philosophers Edmund Burke, Immanuel Kant, and Arthur Schopenhauer developed their aesthetic theories. This was the age we call the Enlightenment – the time when human consciousness moved to prioritize the rational human mind as the prime mover in defining the world. It was a time of high confidence in the powers of the human being as the apex of experience – the belief in mind over matter, the human over nature. The Enlightenment is our inheritance, and it is the worldview that shapes how we approach and understand all reality even today, including art and the natural world.

The response to the natural world took a sharp turn during the Enlightenment. Prior to the time of the Enlightenment, nature was perceived as having agency – the power and even the intent to act independently of, and counter to, human will. In Enlightenment thought, the positions were reversed; humankind now ruled. Rational thought superseded religions, mythology, mysticism and paganism, replacing them with rationalism, empiricism and the scientific method. The Enlightenment worldview undergirded the next two centuries of human so-called progress: the industrial revolution, advances in medicine, the scientific study of nature – all reflecting the superiority of humanity over the natural world.

It is our cultural mindset today, but it is increasingly being called into question as its effects have altered the natural world and placed humankind in jeopardy.

By the middle of the twentieth century, the Enlightenment worldview was challenged on all fronts. The devastation wrought by two world wars and the Holocaust undermined the confidence in human rationalism. One of the principle critical theorists of the Frankfort School, Theodor Adorno, took up a critique of Enlightenment thought and a reassessment of aesthetic theory, calling into question Immanuel Kant's idea of reflection from a safe distance. Adorno collapsed that distance, putting humankind and nature once again in juxtaposition.

The next major shift in aesthetics, for the purposes of my argument, came with the contemporary writings of environmental philosopher Timothy Morton, the ethical arguments of Emmanuel Levinas, and the cultural configurations of German philosopher Peter Sloterdijk. Not only do we have to reject the distances we set up between humans and the environment, we need to recall and revitalize the pre-modern intimate and interconnected relationship with the natural world.

Both senses of the sublime – the experience of awe and immensity, of something overwhelming, beyond our comprehension; and the experience of a profound fear in the presence of a great and unpredictable force of nature – can shape our response to the Anthropocene. The challenges of the Anthropocene have opened up the opportunity to reconsider the dark side of the sublime and our own fears, as we come face to face with the agency of nature. Instead of a rational confidence, once again we experience the raw, primal fear of the consequences of nature's indifferent, and potentially life-threatening force.

The sublime is more than my personal experience on the seashore. The experience of the sublime is the beating heart of all art. Very simply, it is how we express our engagement with something beyond our ordinary comprehension. It is not an emotional experience, or a rational one – the experience of the sublime is its own aesthetic affect – it touches us at the feeling and experiencing core of our humanity.

Examining the sublime is crucial to understanding and articulating our aesthetic responses to the Anthropocene for several reasons:

- It explains the power of art to generate a strong, visceral level of knowing;
- It frames the interactivity between intellect, our emotions, and our affects;
- Studying the changes in the concept over the centuries can help us develop a new aesthetic that expresses a more responsible relationship with the natural world; and finally,
- Through the sublime, we can better appreciate the complexity and power of the Anthropocene.

Experiencing the sublime demands our attention, waking us up to the challenges ahead of us. Human beings are not very good at addressing long term challenges – we're wired for living in the present, focusing on short term outcomes, and instant gratification. Sometimes we need a shock to overcome complacency. We also have a strong sense of self-preservation and the potential for using our creativity, mental skills, and technological skills to respond to threats. In the face of the shock and awe induced by the Anthropocene, we are confronted with the choice of fight or flight. Since we first fully realized the implications of the climate crisis for human survival, we have tried flight – in effect, global warming denial. The second option, fight – ACTION – is the only realistic option left. We have to respond to the dire threats of the Anthropocene. The experience of the sublime, I will argue, is at the heart of our dilemma and our salvation. Awakened through this powerful aesthetic experience, we can act on our strong sense of self-preservation to respond.

A new sublime can be the key, I believe, to cracking open the spectacle of alienation we experience in the Kantian relationship with nature. Our challenge is to transform fear into action – and art can be the means to break through the veil that hides the awareness of the Anthropocene. We stand enthralled in front of a distant, but uncanny awareness that our world has changed dramatically. We have been asleep. We now know that the wizard behind the curtain who manipulated our delusions about the destruction of the planet has been exposed, and we are confronted with the consequences of our own actions. We have been living in a delusion of our own making – it's time to put on the glasses discovered in the film *They Live*, and see the truth.

The Enlightenment Sublime
Overcoming Primal Terror with Reason

Irish philosopher and Member of Parliament Edmund Burke is remembered primarily as a conservative political theorist, but, as a young man, he also made important contributions to the budding field of aesthetics with the publication of *A Philosophical Enquiry into the Origin of Our Ideas of the Sublime and the Beautiful* in 1757. According to Burke, the sublime reflects our response to terror: "Whatever is fitted in any sort to excite the ideas of pain, and danger, that is to say, whatever is in any sort terrible, or is conversant about terrible objects, or operates in a manner analogous to terror, is a source of the sublime; that is, it is productive of the strongest emotion which the mind is capable of feeling."[2]

Burke linked the human experience of the sublime to our relationship with nature. His formulation of the sublime centered on the primordial experience of early humans living in fear and terror, constantly subject to the power of nature, in an era when only a thin veil separated human awareness and agency from the environment. In the face of the overwhelming forces of nature, the human instinct for self-preservation produced a fearful response. Burke used the metaphor of a turbulent sea to convey this deep feeling of terror: "[T]he ocean is an object of no small terror. Indeed terror is in all cases whatsoever, either openly or latently the ruling principle of the sublime."[3] The very real threat of the violent phenomena of nature provided the context for experiencing the sublime.

Following on Burke's early theorizations, German philosopher Immanuel Kant recognized the power of fear, and to mitigate it, he granted a central role in his aesthetics to the power of human reason to overcome the psychological affects of fear and terror. Reason afforded an alternative response to fear – it promised disinterestedness and objectivity, a safe harbor away from the force of nature. The cool, rational, and aloof observer could experience terror at arm's length. Agency shifted from an all-powerful nature to the human mind, the human subject now in control of nature objectified. Kant's sublime reflected the *Weltanschauung* of the Enlightenment, a world view reflecting humankind's increasing sense of confidence and power over nature.

While Kant's conceptualization of the "dynamical sublime" is characterized by a powerful fear of total annihilation, this threat is mitigated by reason which tells us that in the face of raw nature we are still free of its power, and we can experience this fear, and our freedom, in safety. Like Burke, Kant used the metaphor of the ocean to describe the direct experience of the sublime. We are not to project our intellectual understanding into describing the sea, but instead it should be apprehended affectively and aesthetically at a distance.

An important aspect of the sublime is the distance created between the viewer and the fear of the agency of violent nature that threatens. Pleasure or delight arise because the viewer experiences the image in safety, distant from the fear portrayed, and thus is free to enjoy nature as awesome power in the abstract.

British painter J. M. W. Turner's early nineteenth century paintings of shipwrecks and violent storms at sea are perhaps the most widely recognized illustrations in western art of the terror of the Kantian sublime and the fearsome power of nature unleashed, but contained by reason. In her study of how shipwreck art articulated the sublime, art historian Christine Riding observed:

> In general terms, the fact that it is a subject that encourages the spectator to imagine 'pain and danger' and 'self-preservation', 'without being actually in such circumstances' may well be why shipwreck, as a potentially life-threatening event, was suited to the sublime in all its various literary and artistic manifestations, and why it was so regularly adopted by artists planning works for public consumption.[4]

Again, we see the sea employed a central metaphor for conveying the overpowering awe of the cosmos, or of a natural phenomenon beyond our grasp or immediate understanding. Similarly, we can point to the contemporary macabre fascination with the horror of the sinking of the Titanic, the existential threat of sharks portrayed so vividly in the film *Jaws*, or the popularity of horror films such as *Godzilla, Aliens,* and Jurassic Park.

The sublime can also contain an element of enjoyment because the viewer/subject can experience a horrific event at a distance, resurrecting the primeval fear while at the same time mitigating it. Commenting on Kant, contemporary philosopher Gene Ray noted, "Kant emphasizes that a necessary condition of the feeling of negative pleasure is the element of safety that protects the subject's encounter with the terrifying power or size of nature."[5] The sublime produced a kind of delight (or a pleasure or sense of relief) achieved when the threat is perceived from a distance.

Another important aspect of the eighteenth-century sublime was the experience of vastness and immensity. Following in Kant's philosophical footsteps, Schopenhauer expanded Kant's characterization of the sublime, and argued that the feeling of the sublime not only reflected the fear of a threat to the life of the observer, but also filled the viewer with awe and conveyed transcendence: "The impression of the sublime may be produced in quite another way, by presenting a mere immensity in space and time; its immeasurable greatness dwindles the individual to nothing."[6]

The emphasis in these Enlightenment notions of the sublime is on the human subject being the center of the "action." The viewer is positioned at a distance, not prey to the horrific power of nature. The human being shapes and defines the reality as nature recedes to being the object of subjective observation and not as an active, all powerful independent agent. There is a clear demarcation between the human being and nature in Burke, Kant, and Schopenhauer, a premise that will be called into question by more contemporary readings of the sublime. As a new, more nuanced sublime evolved, the supremacy of humankind and the distance from the world eroded as nature moved to the forefront of a dialectical action – a participating, active agent, not mere backdrop for reflection, or stimulus for pleasure from a distance.

The Kantian sublime is a kind of spectacle that continues to shape our contemporary ways of looking at the world. It is an abstract experience, rather than a real experience. I sit in my living room watching the images of fires burning the hillsides of California. I am safe, I can contemplate the horror, but not be affected by it. I am the arrogant one, the watcher. Over and over California victims displaced by the wildfires commented that it was "like watching a movie" to see the flames approaching their neighborhoods, consuming their homes. Their minds could not accept and process what was happening. In the same way, reporters interviewing neighbors where a murder has occurred, hears a similar message: "This is a nice neighborhood, nothing like this has ever happened here before." Both examples suggest that we accept our role as observers, and as observers we do not have to participate in the horrors in front of us.

Theodor Adorno
The Shudder as Aesthetic Awakening

Little changed in the interpretation and use of the term sublime from the eighteenth century until the twentieth century. In *Dialectics of Enlightenment*, Max Horkheimer and Theodor Adorno undertook a broad critique of Enlightenment thinking. In the course of their analysis, they revisited the idea of the sublime. In their telling, the relationship between humans and nature changed across the millennia. Early in human history, there was a seamless unity between humans and nature – a state Adorno calls *mana*. In the state of *mana*, for example, early artists created images of animals that were understood as real, not representations – and had real powers. Adorno likened the aesthetic experience of *mana* to the sublime, calling it the "shudder" – a concept central to his understanding of the primitive human encounter with the natural world and to the contemporary sublime. The shudder is the recognition and awareness of the sublime. Confronted with the unknown in the natural world which induced fear, primitive humans sought to understand and express their connections with the natural world by accepting the power and presence of an indifferent nature. Enlightenment thought, with its separation of subject and object and its prioritizing of the human subject, led to our disenchantment from nature. To re-enchant nature, to reconnect with the primitive consciousness where nature and human were one interrelated being, it was necessary to critically challenge and decenter the role of reason which had led to the notion of human domination of nature. In Adorno and Horkheimer's reformulation of the sublime, nature is no longer the object of rational reflection, and the distance between the Kantian subject and the object is collapsed.

Philosophies of the Enlightenment created a rupture with primitive ways of knowing, severing the human connection with nature: "The disenchantment of the world is the extirpation of animism. ... The moist, the indivisible, air, and fire [the elements], which they [the pre-Socratic cosmologies] hold to be the primal matter of nature, are already rationalizations of the mythic mode of apprehension."[7] With the supercession of rational thought, "the fear of uncomprehended, threatening nature, the consequence of its very materialization and objectification, was reduced to animistic superstition, and the subjugation of nature was made the absolute purpose of life within and without."[8] Adorno neatly summarized the transition from the ancient close relationship between humans and nature, to the Enlightenment worldview where nature becomes an object: "Myth turns into enlightenment, and nature into mere objectivity."[9]

In his critique of Enlightenment thought, Adorno argued that *mana* could be recovered and the world can be re-enchanted. He concluded that the primitive awareness of the unknown and of the power of nature lived on in art, reconnecting humanity with nature unmediated by rational thought – a return to the magic of the world:

> The work of art still has something in common with enchantment: it posits its own, self-enclosed area, which is withdrawn from the context of profane existence, and in which special laws apply. ... It is in the nature of the work of art, or aesthetic semblance, to be what the new, terrifying occurrence became in the primitive's magic: the appearance of the whole in the particular. ... This constitutes its aura.[10]

True art, he argued, recuperated the primeval human relationship with nature through the experience of the shudder. The shudder is that sense of recognition, the affective experience of the sublime.[11] Artworks that actualize the shudder create a response in the viewer (the subject), grabbing their attention. The shudder is not an emotional response, but a transactional, interactive affect: "The shock aroused by important works is not employed to trigger personal, otherwise repressed emotions. Rather this shock is the moment in which recipients forget themselves and disappear into the work; it is the moment of being shaken. The recipients lose their footing; the possibility of truth, embodied in the aesthetic image, becomes tangible."[12] We are returned to a primordial awareness of repressed nature where nature resumes its role and enmeshed agency in the lives of humans. Adorno recalls and revitalizes, through the experience of the sublime in art, the ancient connection with nature.

Adorno's formulation of the sublime removes the protection granted by distance in Kant's rendering of the sublime. Instead, the subject is enmeshed and vulnerable to the original experience of terror in the face of nature's power. In her assessment of aesthetics in the Anthropocene, contemporary cultural theorist Marah Nagelhout called for a renewed appreciation of the catastrophic, basing her argument on a careful reading of Adorno.[13] Sensitivity to the catastrophic powers of nature is central to any considered response to the Anthropocene. As we come to see the increasing unpredictability of natural forces in the Anthropocene, the distance between humans and nature disappears and the ancient terror is again manifested.

Aesthetics in the Anthropocene must articulate a new sublime reflecting the catastrophic, apocalyptic powers of nature that characterized the early human relationship with nature. The terror of the unknown, the fear of the power of nature is once again front and center. Nature can no longer be background to humanity. The Anthropocene recognizes the agency of both human beings and other beings. Adorno's "shudder" is elicited by the fears embedded in the concept of the Anthropocene. Nagelhout concluded: "For it is when artworks conjure up the image of catastrophe, 'an image that is not a copy of the event but a cipher of its potential' (Adorno 1997, 33), that artworks expand the cognitive realm of possibilities."[14] As will also be shown in a later chapter, images of the agency of the sea have played an important role in bringing to life the primitive experience of the catastrophic when they are embedded in the narrative of the Anthropocene.

Adorno provided a pathway toward a new aesthetic of the Anthropocene that reconnects human beings with the natural world. We no longer have the luxury of the Kantian safe harbor of objectivity from which to experience the unfolding disaster confronting us. Nagelhout was interested in finding "[A]n aesthetic theory that adopts the paradigmatic shift in consciousness the climate crisis demands of us," arguing that "With the creeping catastrophism of the climate crisis, fear of nature in what would be a sublime moment – by Kant's formulation – is appended not by the promise of safety, but the imminence of our total demise. … [N]ature has become a threatening entity that looms over our existence."[15] Adorno's sublime provides important insights into how human beings might potentially address the coming crisis: "In the age of the Anthropocene, our vulnerability cannot be avoided, and it is perhaps only at this critical juncture in our history that the sublime can return to nature."[16] Nagelhout concluded that nature must now be reckoned with as a force in its own right – a force we cannot experience at a distance, but must embrace: "Indeed, the climate crisis shows just how unsuitable nature is to human understanding … For this reason, nature as chaos, as an arbitrator of agency capable of defining our collective future, denounces the reconcilability of nature and man by human reason."[17]

As we have seen, the idea of the sublime, from the perspective of the human subject, changed dramatically from the seventeenth century insights of Burke and Kant to Adorno's critique in the twentieth century. Enlightenment thinking made it impossible for human beings to fully realize a democratic and dialectical relationship between subject and object. By the twentieth century, the fear and terror of nature had faded; the perception of the threat of annihilation had subsided as nature was suppressed by the sheer force and power of human domination. This transition paralleled the movement from a world of myth and magic to the modern world of rational thought and faith in science. Nature had come under increasing human control; the seas had been conquered; all the continents had been discovered and colonized; and the industrial revolution transcended many natural powers as coal and fossil fuels expanded productive force exponentially.

Adorno called all of this into question. All was not well. The horrors of Hiroshima, Nagasaki, and the Holocaust reminded us that the terrors persisted despite our hubris. We realized we had created our own horrors and there was no escaping the consequences. By the middle of the twentieth century, it was clear that there were major fault lines in the human/natural world relationship. Pollution from fossil fuel industrial processes threatened air and water quality; chemical poisons were killing wildlife and humans both; tinkering with nuclear energy spread radiation around the globe; and the exploitation of natural resources to fuel consumer capitalism expanded exponentially.

Timothy Morton
Hyperobjects and Embracing the Immensity

I'd like to turn now to a contemporary ecological philosopher, Timothy Morton. I found his work challenging, but incredibly illuminating as we scramble for a new way to understand the changes our world is undergoing. Morton rejected the anthropocentrism of Enlightenment thinking and instead, embraces the idea of co-equality between all beings. In Being Ecological, Morton argued that by removing humanity from the archaic notion that we are the center of the universe, humanity is freed up to understand our place in the world in a new manner. Humans are no longer the center of the world, we are not the definers of reality, but merely one player amongst many. In a light-hearted observation in Hyperobjects, Morton observed that this new way of looking at reality "radically displaces the human by insisting that my being is not everything it's cracked up to be – or rather that the being of a paper cup is as profound as mine."18

I was especially fascinated by Morton's concept of the hyperobject – his way of redefining how we have to begin to comprehend the new environmental challenges confronting humankind in a new way of thinking. According to Morton, hyperobjects will change the very way we think about our world and our place in it. So what are hyperobjects? Hyperobjects are unfathomable, they are complex, and they engender our awe. Hyperobjects are not visible; for example, we can't "see" the hyperobjects of global warming or radiation in real, human time; we can only finally see their effects over geologic time and in changes to our genetic makeup or in the suffering and death of people exposed to its invisible effects. Hyperobjects are big, and as in the case of global warming, they can be distributed across geologic time and space.

Our growing awareness of the hyperobject of global warming demands a change in our interactions with the natural world. It is impossible, Morton argued, for us to fully grasp the hyperobject "global warming." We are part of that hyperobject, not outside of it. This makes it very difficult for human beings to comprehend, let alone respond:

> [Y]ou are never directly experiencing global warming as such. Nowhere in the long list of catastrophic weather events – which will increase as global warming takes off – will you find global warming. But global warming is as real as this sentence. ... How can we account for this? By arguing that global warming,

like all hyperobjects, is nonlocal: it's massively distributed in time and space.[19]

We are, Morton argued, "caught in a trap of hyperobjects," meaning that we are glued to our reality. Hyperobjects "stick to us … are us." He offered radioactive materials as an example of what he called viscosity: "The more you try to get rid of them, the more you realize you can't get rid of them. They seriously undermine the notion of 'away.'"[20] There is no safe harbor. According to Morton, "[H]yperobjects are here, like the ghosts in *The Sixth Sense*."[21] Morton used another example of radiation to explain just how the viscosity of hyperobjects affects us: "We know we are bathed in alpha, beta, and gamma rays emanating from the dust particles that now span the globe. These particles coexist with us. They are not part of some enormous bowl called Nature; they are beings like us, strange strangers. … There is no exit from this situation."[22]

Because we are co-equal with the global warming object, we will have to develop a relationship with these mysterious objects that carry their own beingness, their own presence – even their own voices. This calls for a radical rethinking of all our preciously held assumptions about human preeminence. According to Morton: "Hyperobjects pose numerous threats to individualism, nationalism, anti-intellectualism, racism, speciesism, anthropocentrism, you name it. Possibly even capitalism itself."[23]

Morton's idea of "hyperobjects" provides a deeper and more nuanced contemporary application of the idea of immensity, and forms the foundations of a new sublime for the Anthropocene. An entirely new understanding about our place in the world comes into being with the change in perspective. Instead of human beings calling the shots, nature becomes an active player.

According to Morton, we no longer have the luxury of the Kantian sublime – the safe distance from which we can contemplate the "oppressive claustrophobic horror" of global warming. Morton echoed Adorno's depiction of the new sublime as intimate, where we are confronted with our vulnerability and our inability to separate from the catastrophe confronting us.

> But inside the belly of the whale that is global warming, it's oppressive and hot and there's no "away" anymore. And it's profoundly regressing: a toxic intrauterine experience, on top of which we must assume responsibility for it. And what neonatal or prenatal infant should be responsible for her mother's existence? Global warming is in the uncanny valley, as far as hyperobjects go.[24]

Morton referenced the "uncanny," a psychological concept formulated by Sigmund Freud in a 1919 essay. On the one hand, the uncanny is a manifestation of an uncertain fear, and at the same time it seems familiar to us. We recognize that global warming raises the hackles on the back of our necks, it is "here," but we can't quite put our finger on it, it is an uncanny experience. This creates anxiety and discomfort. In the passage below, Morton also seems to bring together Freud's uncanny sense of familiarity with Adorno's notion of a sublime where aesthetic distance is evaporated and we stand face-to-face with the Other, recognizing our vulnerability in this world that we have damaged. It is here that we once again confront Adorno's shudder:

> Recognition of the uncanny nonhuman must by definition first consist of a terrifying glimpse of ghosts, a glimpse that makes one's physicality resonate (suggesting the Latin horreo, I bristle): as Adorno says, the primordial aesthetic experience is goose bumps. Yet this is precisely the aesthetic of the hyperobject, which can only be detected as a ghostly spectrality that comes in and out of phase with normalized human spacetime.[25]

In *Realistic Magic*, Morton outlined a new understanding of the sublime that demands engagement, a real intimacy, an encounter and a recognition of the relationship with the other phenomena and beings in our shared world:

> What we need is a more speculative sublime that actually tries to become intimate with the other … This is precisely the kind of intimacy prohibited by Kant, in which the sublime requires a Goldilocks aesthetic distance, not too close and not too far away. … I shall now argue for a speculative sublime, an object-oriented sublime to be more precise. There is a model for just such a sublime on the market – the oldest extant text on the sublime, *Peri Hypsous* by Longinus. … The Longinian sublime is based on coexistence. At least one other thing exists, apart from me: that "noble mind," whose footprint I find in my inner space. … This is good news in an ecological era. Before it's fear or freedom [Burke or Kant], the sublime is coexistence.[26]

The New Sublime
Being with the Other

What has to change is how we think about our relationship with the natural world and the other beings with whom we share the world. Contemporary philosopher and Talmudic scholar Emmanuel Levinas is cited (often only in passing, which is unfortunate) by several philosophers writing about the

Anthropocene. Levinas proposed an entirely new approach to ethics based on a principle of reciprocal obligation – paying attention to an Other. Ethics is central to developing new ground rules for interacting with the planet, its beings, even its hyperobjects.

Levinas' basic argument is that human beings are ethically locked together in a binding relationship of responsibility. And even more significantly, he stated that it is the Other that determines the individual's ethical responsibility, not the subjective self: "I understand responsibility as responsibility for the Other, thus as responsibility for what is not my deed, or for what does not even matter to me; or which precisely does matter to me, is met by me as face."[27] While his work is focused on human relationships, I think his approach to ethics has broad application to understanding the interactions between humankind and the natural world. I would ask the reader to adapt Levinas' ideas to our exploration of the Anthropocene, substituting the relationship between humankind and nature for his descriptions of the interactions between two individuals. The Other, for our purposes, is the natural world, and the foundation of the new sublime is grounded in ethics.

"Face" as Levinas used the term, is the experience of a genuine encounter; it is an interactive relationship built on responsiveness of the self to the Other, of being *for* the Other. The face is an encounter with the infinite – even with a hyperobject we might speculate. According to Australian philosopher David Hickling, Levinas' "face" is not an actual face, but an expression of the experience of the other:

> When I encounter an Other, what is signified is not another "me" (which would be, for Levinas, 'sympathy') but something Other, something that I cannot understand or comprehend – as mentioned above it is infinite. It is alien, it is an alterity. The Face is the locus of the ethical relationship with the Other. ... Once we are in a face-to-face relationship with the Other, ... then we are commanded by the Other. We are, in effect, subject to the Other. We are incapable of indifference.[28]

It is the Other who determines the nature of the ethical relationship. Levinas' ethics is up close and personal, a binding relationship; intimacy and obligation. The Other is not an object, but another participating being, and the obligation to respond defines the relationship. According to Thai philosopher Kajornpat Tangyin, Levinas' position is a very radical one: "He wants philosophy to begin with this relation, and this relation comes with an ethical demand, i.e., before the face of the other you shall not kill and in fact, you have to defend the life of the other."[29]

In an interview with philosopher Richard Kearney, Levinas further defined this relationship of obligation and responsibility: "The ethical 'I' is subjectivity precisely in so far as it kneels before the other, sacrificing its own liberty to the more primordial call of the other. For me, the freedom of the subject is not the highest or primary value. ... As soon as I acknowledge that it is 'I' who am responsible, I accept that my freedom is anteceded by an obligation to the other."[30]

Furthermore, according to Tangyin, there is no expectation of reciprocity: "For Levinas, the asymmetry of the ethical relationship is very important for human relationships. It does not imply demanding the other's responsibility for me; my responsibility for the other does not mean the other will do the same in return."[31] In short, according to German philosopher Peter Sloterdijk, the Other "is the one to whom one always owes something."[32] Aesthetics, then, becomes the experience of this ethic of responsibility, of desiring to be responsive to the Other, to care for the Other because to be myself, I must fulfill this obligation.

Listening to the Other

How, then, do we embrace the wild and respond ethically to nature? I think it begins with a caring for the Other/nature, a respect for the Other/nature, and a desire to interact. To live we must respect the aliveness of all beings. In a loving passage from the poem "Song of Myself," American poet Walt Whitman first reaches out to the sea, resigning himself to its power and force, and then he and the sea embrace in an intimate, sensuous manner. The two are one in a responsive sharing of experience.

> You sea! I resign myself to you also – I guess what you mean,
> I behold from the beach your crooked fingers,
> I believe you refuse to go back without feeling of me,
> We must have a turn together, I undress, hurry me out of sight of the land,
> Cushion me soft, rock me in billowy drowse,
> Dash me with amorous wet, I can repay you.
>
> Sea of stretch'd ground-swells,
> Sea breathing broad and convulsive breaths,
> Sea of the brine of life and of unshovell'd yet always-ready graves,
> Howler and scooper of storms, capricious and dainty sea,
> I am integral with you, I too am of one phase and of all phases.[33]

As awareness of our interconnectedness with the earth expands, we are confronted with the ethical imperative to reconnect with nature. In the Anthropocene, humanity can no longer assume the dominant role in the discourse. The first challenge we face is sensitizing ourselves to the interconnectivity of humans with the planet – listening to the Others. According to author Heather Davis and philosopher Etienne Turpin: "If we are to learn to adapt in this world, we will need to do so with all the other creatures; seeing from their perspective is central to re-organizing our knowledge and perceptions."[34]

In a similar vein, American philosopher Arnold Berleant called for an aesthetic that is descriptive of a participatory environment, or what he calls "aesthetic engagement" – an environment that is "a field in which there is a reciprocal action of organism on environment and environment on organism, and in which there is no sharp demarcation between them. Such a pattern may be thought of as a participatory model of experience."[35] No longer the neutral observer, we are active participants in a community of related beings:

> We are implicated in a constant process of action and response from which it is not possible to stand apart. A physical interaction of body and setting, a psychological interconnection of consciousness and culture, a dynamic harmony of sensory awareness all make a person inseparable from his or her environmental situation. Traditional dualisms, such as those separating idea and object, self and others, inner consciousness and external world, all dissolve in the integration of person and place.[36]

In *The Re-enchantment of the World,* historian Morris Berman eloquently described the outlines of this new relationship, recognizing our enmeshment.

> In pitting his own survival against the survival of the rest of the eco-system, in adopting the Baconian program of technological mastery, Western man has managed, in a mere three centuries, to throw his own survival into question. The true unit of survival, and of Mind, is not organism or species, but organism + environment, species + environment. If you choose the wrong unit, and believe it is somehow all right to pollute Lake Erie until it loses its Mind, then you will go a little insane yourself, because you are a sub-Mind in a larger Mind that you have driven a bit crazy. ... [T]here are clear limits to how many times you can create such situations before the planet decides to render you extinct in order to save itself.[37]

We can begin by being especially responsive to the damages caused by human activity – the cries of soils wearing out, oceans groaning under the weight of pollution and rising temperatures, rivers unable to cleanse themselves any longer, the composition of air affecting the growth of plant life.

It will not be easy to establish these new relationships and obligations, as Alison Stone argued: "To 're-enchant' nature, conversely, would be to find in it a meaning that cannot be fully understood."[38] And I would add, may never be understood. We cannot define the Other, the Other has its own presence and its own meaning to which we can only be attentive. According to Adorno, Stone notes, "living beings *suffer (leiden)* from having their spontaneous tendencies thwarted."[39] Is it possible, Stone asked, to criticize what is our dominant worldview of destruction and exploitation? The goal, she continued, should be to "motivate us to criticize and change our relationship with nature, and to feel guilt for having damaged it and to seek to make amends."[40] The damage we have done to the earth, is ultimately damage done to ourselves. The challenge then is to move beyond mourning the damage and the losses and accept responsibility to make things right.

Human beings are no longer understood to be the sole actors and directors in a world of hyperobjects – only one actor among many. Agency is a power of all beings, however unpredictable and erratic or however active or quiescent it may seem. We are finally coming to the awareness that everything has agency, and we're no longer in charge. Nature has its own trajectory that we must learn to adapt to and accommodate. Every being seeks to live, and it is in recognizing our kinship at this level with the planet that we can experience a solidarity of being. Artist Ricardo Levins Morales, in an essay entitled "Resilience: Another Name for Life," reminded us of the majestic power of the life force, and the passions underlying all life to achieve the richest beingness:

> Resilience is the central story of life on Earth. It could be considered another word for the life force. The never-ending dance of adaptation and innovation keeps species, communities and individuals constantly adjusting to micro changes in their environments as well as major ones.
>
> This is not to minimize the cascade of crises facing our world. It's just that no one knows the limits of resilience on a global scale. Global processes are not just bigger versions of small scale ones. Each level of Earth's systems dances to its own beat. Given that we know so little we must do what all life does in the face of uncertainty: innovate, adapt, resist, respond, transform.[41]

All of nature is resourceful and seeks to realize its full life potential. New bacteria overcome the effects of human antibiotics (anti-biotics = anti-the bios); weeds transform their DNA to resist Monsanto's (now Bayer) herbicides; rivers dammed and forced into new, unstable channels suddenly burst forth and reclaim forgotten fields. This passion for life includes humankind and we will rise to the challenge of saving ourselves along with all the other beings.

What is required, according to Professor Patti Pente, is a major shift in reality to reflect our new relationship. "Specifically, as posthumans, we are as much influenced and dynamically formed by the materials of our environment as we influence them. Thus, posthuman theory suggests *becoming with* the world."[42] Ontological change is necessary as we adapt to this new reality "to help people shift from seeing themselves as human who use/save the planet to entities who are symbiotic parts of the planet, sharing this existence with many equally integrated non/inhuman entities."[43]

As we come to understand that human beings are having a major impact in geologic time, we also have to consider that other beings and natural phenomena have a similar power of acting. It is no longer enough to just recognize what we are doing to the planet, it may also be a matter of respecting and incorporating into our thinking what the other actors on the planet are capable of doing to us. Novelist Amitav Ghosh speculated that human beings are now experiencing the re-emergence of a sense of the uncanny in response to global warming.

> Yet now our gaze seems to be turning again; the uncanny and improbable events that are beating at our doors seem to have stirred a sense of recognition, an awareness that humans were never alone, that we have always been surrounded by beings of all sorts who share elements of that which we had thought to be more distinctively our own: the capacities of will, thought and consciousness.[44]

The veneer of civilization is, after all, very thin, and we are on the verge of recouping the sublime sense of awe and dread – only this time out of a sense of responsibility and respect. The new sublime recognizes that there is agency in nature, a force beyond our control. As we become more attentive to the agency of the planet and our interactions with the natural world, our responses cannot be individual. Without a broader awareness of our impact as a species, we cannot address the looming problems. We need to expand thinking of how humans as a species interrelate with the natural world. This requires, of course, that individuals think of themselves as members of the human species, of a collective.

It also goes without saying that the challenge to develop appreciation of the Other must include our relationships with those we have identified as "other" within the human family itself – whether by race or nationality, by ability, gender, or other arbitrary divisions. We need to overcome our differences in order to survive because we all are in this together. We need one another to address the catastrophe to save the world and ourselves.

Our obligations to nature will require that we make amends for damages that we have caused. Recognizing the suffering we have inflicted on the planet is the beginning of the healing and restitution process. But it is not just a matter of awareness; it is also a matter of recognizing the urgency of looming catastrophe, accepting the errors made, grieving those decisions, and then anticipating what actions must be taken to remediate, restore, or adapt to the damages already done.

Conclusion

And so we come again to our beginnings, back to a time where we find ourselves in the state of *mana,* back to the conscious intermingling of our being with the powerful forces of the natural world, a world of enchantment. The trees are alive, the animals speak to us. The hunter thanks the beast for its gift of food. Over the past three centuries, our journey as a species away from this enchanted world was a long and alienating one. We moved away from this mutual awareness, instead finding gods, and myths, and magicians to explain the mysterious in nature instead of embracing it and engaging with it. With the discoveries of the so-called scientific revolution, we thought we had found a new, more objective way to transcend our deep fears and terrors about the world around us. And we wanted control. We patted ourselves on the back and explained our new superiority as we embraced the Kantian sublime. The center of Kant's Enlightenment thought did not hold, however, as nature returned in a geological moment's time to show us our limits once again – the Anthropocene. We now know that all of nature has its own agency; every ecology, every microbiology has its own way of being in the world. Our task, now (with a sense of profound urgency), is to develop a renewed relationship, and learn to coexist with the other objects in our universe.

Now that we see the spectacle of the Enlightenment as the unreal dream, we can begin to deconstruct those thought patterns and visions that we actualized into bringing the world to the brink of destruction. The optimism and the hubris of the Enlightenment, we now see, was merely an intellectual illusion, a false fantasy. This is what philosophy does – shows us how we think, pointing us to new horizons. We have a stark choice to make between two constructions of reality – the petro reality of modernity or the survival reality of the future.

MAKING ART ON A DYING PLANET
IMAGINING THE ANTHROPOCENE

> Unjack your cortex fully from the fear-soaked narratives
> of insanity, and let the true beauty of our real world flood
> your senses. Let the grief of what we have unknowingly
> done send you crashing to your knees in sorrow. And
> when you're ready, stand up. We have much work to do.
> – Caitlin Johnstone[1]

As the chaos and uncertainty around global warming increases, and as the reality of the Anthropocene surrounds us, it will not be easy for artists to harness the powers of creation in the political fight for the planet. Surrounded by fellow human beings filled with dread and confusion, the planet ravaged by fires and floods, ecological niches crumbling, the extinction of plants and animals – ah, yes, the pain and suffering and hunger and fear will call for clear visions and compassionate hearts. How do artists address the looming crisis confronting humanity? What does art look like as the Anthropocene unfolds in this time of transition and uncertainty? Of impending suffering, of potential catastrophe already underway?

How artists can address the very real fears of the damaged present, and at the same time, communicate a message of hope and the possibility of a new, but different future, will be the focus of this chapter. All artistic skills and knowledge will be called upon as we go forward. I offer a series of broad observations but the visions and the symbols, the colors and shapes and forms, and the feelings and emotions will ignite in the artistic imaginations of individual artists.

Art in the Anthropocene must tell stories, stories describing our relationship to the natural world, stories that will organize the challenges confronting the world into meaningful messages. Art will be shaped by a deep knowledge of the shared stories of the human/earth relationship – how our politics of domination and exploitation have harmed the earth, how the earth has suffered at our hand, how our thinking is changing/must change; and how we can move forward to mitigate the damage we have done.

Art will incorporate a new, *more meaningful experience of the sublime* that articulates an awareness of the threat to human survival, our fears for the future, and the powerful life force for self-preservation embedded in all beings. By facing our deepest fears and anxieties, coming to terms with our vulnerability in the face of impending death and catastrophe, humanity can move beyond denial and paralysis to awareness and action. The new sublime will open our eyes to the story we are currently living within, and how to transcend it.

Art will be a political art, designed specifically to *transform thinking and action*. It will call out for action, not merely reflection. In that sense, it will share the techniques and objectives of propaganda.

Art will recognize that *we exist in the interlocking network* of the earth. The world is not one ecology or one being, but many ecologies and beings all interconnected in ways we are just beginning to comprehend. Complexity lies at the heart of nature and in the human/nature relationship and the beauty of these intricacies should foreground our vision of future connectedness. Over time, humankind has abandoned its former intimacy with the natural world. This now needs to be reclaimed. Reclaiming this intimacy will be reflected in a new aesthetic of relationship. There is no independent, objective "nature" out there; there is only human/nature/we – partners in the work of mutual survival.

Art will embrace our interconnectedness and reflect this belonging. It will be rooted in a connection to our membership in the hyperobject of the Anthropocene, in a global human/earth community, and in a local human/earth community, addressing how the changing earth affects us here and now. It will reflect both the immensity of planetary awareness, and the familiar connections with the local neighborhood. It will seek to articulate shared cultural meaning and build relationships between all beings; and its principles will be accessible across cultures.

A deep knowledge of the natural world is a must for artists of the Anthropocene. Chinese philosopher of aesthetics Xiangzhan Cheng argued that it was critical that humanity understand scientific knowledge and rely "on the ecological knowledge to refine taste and to enjoy the hidden rich aesthetic properties of the ordinary (even the trivial)."[2] According to environmental artist Jennifer Rae, the Center for Research on Environmental Decisions offered several suggestions for connecting people to the challenges of art in the Anthropocene: "In order for the public to fully absorb climate science research, 'it must be actively communicated with appropriate language, metaphor, and analogy; combined with narrative storytelling; made vivid through visual imagery and experiential scenarios; balanced with scientific information; and delivered through trusted messengers in group settings.'"[3] Artists need to have a keen awareness of intimate places where the richness of life can be discovered, as well as apprehend the workings of the whole planet.

Art will honor and articulate *agency for all beings*. All beings have a passion to act on their own behalf, to live their own lives in their own manner. Human beings are no longer in charge. We must hear and respect the other beings of the earth, especially their desire to live their lives to the fullest, which is parallel to our own passion for living. As Gómez-Barris concluded, if we are serious about responding to the crisis confronting us, we must remain open to the myriad connections linking the web of life and allow for "the letting in of the surround as key to imagining life otherwise."[4]

Art will recognize the *power of affect* as reflected in the sublime in reaching people. The very complexity of the problems confronting the people of the earth calls for a lexicon that will resonate at the level of feeling, rather than the emotional, or the rational/scientific level. The art of the Anthropocene does not just need facts, it needs to have an impact at the gut level to jar the awareness, to awaken, to move to action.

Finally, art in the Anthropocene must seek to articulate a message that is both *restorative and regenerative*. It will suggest alternative possibilities and will weave the new stories of new ways to live on the earth in the future.

The Importance of Narrative

What about the stories that are visions of the world, visions of reality that hold people inside, inside the reality of that world, inside their minds that see the world, and in society among each other? What about those stories that tell you how to live and how to stay alive and how to die? Which are the stories you take inside your mind to live by and to create a world, to teach your children, to wake to in the morning, to face each day? Some say those aren't just stories; others

say that's all they are, just stories, and that's plenty enough.
– Karen Tei Yamashita

How do artists tell a new story of the nature/humanity relationship? Narrative is important to storytelling because it organizes information and thought in a sharable package. A narrative distills and simplifies reality, making meaning accessible and portable. A nature narrative should also be grounded in real human experiences with the natural world to reinforce the necessity of mutual survival.

The development of a meaningful narrative to increase awareness and engage reflection is central to art in the Anthropocene, just as it has been throughout history, from the cave paintings to graffiti on neighborhood walls. This narrative will begin with the story of the need to change the relationship of humanity to the planet. It will advocate strong warnings of a potential apocalypse. The icons and symbols and memes and metaphors will be designed to change things, to tell a story that resonates with many people and empowers them to act.

The planet will become the co-narrator, relating the story in real time, as opposed to mimesis, creating of a representation of the story. Professors Tobias Boes and Kath Marshall suggest that "The challenge for Anthropocenic art will clearly be to move ... into a new realm of 'ecodiegesis' that gives a voice to the planet itself."[5] Nature will be there, for example, to remind us of the contradictions between our consumer lifestyles and the natural limits to the resources of the planet. Humankind will no longer be what Timothy Morton referred to as "ennoblers," but active, attentive partners. What is the planet saying? What stories is the earth rendering? According to Morton, we are already beginning to learn. If we listen carefully, an awareness of our role in the hyperobject of global warming and our participation in it is the beginning of our listening and learning, and our recognition of the agency of nature.

> Nonhuman beings are responsible for the next moment of human history and thinking. It is not simply that humans became aware of nonhumans, or that they decided to ennoble some of them by granting them a higher status – or cut themselves down by taking away the status of the human. ... The reality is that hyperobjects were already here, and slowly but surely we understood what they were already saying. They contacted us.[6]

Narrative, of course, is central to conveying a message and to ascribe meaning in all art forms – whether in literary productions, poetry, visual art, music, or film. In *The Great Derangement*, Amitav Ghosh examined how words and images became separated over time, and suggested that the visual image may be the preeminent art form in the Anthropocene. Ghosh noted that while earlier books included images to communicate content, over time, words alone came to dominate the page: "Merely to trace the evolution of the printed book is to observe the slow but inexorable excision of all the pictorial elements ... illuminated borders, portraits, coloring, line drawings, and so on."[7] He also studied the role of the novel in shaping contemporary cultural and political environmental awareness, concluding that in general the novel has failed in this regard. While surrealism, magical realism, and science fiction succeeded in raising environmental awareness to some extent, these have been minor genres. Citing the failure of literature to address the challenges of global warming, Ghosh asked, "Would it follow, then, ... that to think about the Anthropocene will be to think in images, that it will require a departure from our accustomed logocentricism?" ... [W]ith the Internet we were suddenly back in a time when text and image could be twinned with as much facility as in an illuminated manuscript."[8] Narrative can easily be embedded in images. What story does the picture tell? Ghosh was hopeful that the internets and the social media network it created could change the dynamic of the discussion about the environment and the role a visual narrative could play in addressing the pressing problems.

The uncertainty about the global warming changes underway calls for a story that will bring the fragments together in a meaningful way to move us beyond paralysis. Stories will help us frame reality by condensing the complexity of the Anthropocene into meaningful messages. As poet Lucy Ives observed,

> [T]here is, I would argue, a notable hunger in American society for the comforts of a narrative. It is a hunger for a species of meaning-making that is not specifically logical (though it may be that too) but which rather provides an account of how *things*, sometimes sentient, sometimes material, get organized across space and time and in relation to one another and sequentially, such that they become *the way things are,* after having been *the way things were.* Sure, narrative can be revelatory and informative, but it can also be reassuring, grounding. ... [M]ore and more people want narrative, a) because they want to know how we got here and, b) because they want to know what to do next.[9]

Even though we cannot completely comprehend the hyperobject of global warming, we long for a way of getting our heads around it. Art can take on this challenge and address our longing for structure and meaning. Our contemporary world distracts us from fulfilling this basic human need for order and coherence. Sequential continuity is also implicit in the narrative form, a continuity that can connect with direct experience. According to artists and authors David Bayles and Ted Orland, an older sense of art linked the creative process to the collective making of social connections and cultural meaning: "art is something you do out in the world, or something you do about the world, or even something you do *for* the world. The need to make art may not stem solely from a need to express who you are, but from a need to complete a relationship with something outside yourself. As a maker of art you are custodian of issues larger than self."[10] A strong narrative provides the foundation for new visions. Narrative thus serves as an inspiration for change.

A narrative provides coherence to art, the steps in a journey that is real and meaningful. A new context is created where both the artist and the viewer are partners in change, traveling to a new land, new economic practices, new beliefs, new politics and new dreams. The narrative also implies that the journey is not just random. Further, there is a conclusion to the journey, a closure that contains what used to be called "the moral of the story." The narrative offers a shared memory of feeling and experience, a re-collecting of knowledge, and a predictable process.

The New Sublime
The Case for Embracing Death and the Catastrophic

The Anthropocene is an unfolding catastrophe for which humanity is responsible. Despite a scientific consensus that we have only a small window of time in which to respond to the existential threat of rising temperatures, there is little political action. The challenge for artists is formidable: how to raise awareness and inspire an urgent response.

A desire for self-preservation lurks behind every discussion of the Anthropocene. Awareness of the impending environmental crisis triggers in us a fear of impending death. Understanding the aesthetic principle of the sublime offers us a way to examine more closely the role of fear in motivating humans to proactively respond to the climate challenges we face. Can we embrace the catastrophic and move to action?

Generating fear and focusing on impending horror may be the artist's best option at this point in time. According to environmentalist Lawrence Buell: "Apocalypse is the single most powerful master metaphor that the contemporary environmental imagination has at its disposal. ... [F]or the rhetoric of apocalypticism implies that the fate of the world hinges on the arousal of the imagination to a sense of crisis."[11] Geographer Kathryn Yusoff and sociologist Jennifer Gabrys agreed, arguing that fear is important to our understanding and acceptance of the realities of the Anthropocene: "Fear and hope about climate change may or may not drive action, but they are a part of how we register and understand environmental change, and in this sense they are instructive to understanding the imaginings not just of futures, but also of modes of adaptation."[12]

There is a concern that fear can also inhibit action. It can fuel denial by avoiding the possibility of death; it can fuel apathy and inaction by leaving us depressed or uncertain about how to respond. There is a concern that an emphasis on catastrophism will create a kind of cultural paralysis. But as philosopher Timothy Morton noted, the crisis is real, and it is ominous, it is immanent, and our very lives are at stake: "What is global warming anyway? The correct answer is that it is *mass extinction*" – an extinction at least as catastrophic as the loss of the dinosaurs."[13] This extinction that we are currently undergoing, we need to remind ourselves, includes humans. Now is not the time to quibble about how serious the situation is, but to raise a general alarm.

Yusoff and Gabrys studied the emerging genre of science fiction after World War II and its role in shaping the narrative of apocalypse: "The science fiction that emerged from modernity and its cultures of catastrophism displayed a distinct fascination with disaster at moments in history when it became easier to imagine the end of the world than alterative futures."[14] They argued that the genre emerged in response to the early awareness of major environmental and cultural threats: chemical pollution; the fears of annihilation during the Cold War; the very real possibility of atomic fallout creating monsters; the emerging threat of overpopulation, among others.

Although the effectiveness of doomsday narratives in motivating people to act has been questioned, Yusoff and Gabrys called for a more subtle reading of the impact of fear on activism. Doomsday narratives have the potential to initiate activities that respond proactively to threats. We should not dismiss "the creative role of fiction and the cautionary offerings of the disaster. A story at its best, asks us to imagine alongside the protagonist of a story the full range of emotional challenges and difficult choices that have to be made."[15]

In *The Uninhabitable Earth*, David Wallace-Wells argued that debating, denying, delaying tactics are no longer sufficient, and urged more direct action. The first sentence in his book – "It is worse, much worse, than you think" – has been called "alarmist," but he defended the need for taking aggressive action to motivate people to action in an interview with the *New York Times*.

> [W]e're at a point where alarmism and catastrophic thinking are valuable, for several reasons. ... The first is that climate change is a crisis precisely because it is a looming catastrophe that demands an aggressive global response, now. In other words, it is right to be alarmed. ... [Second] By defining the "boundaries of conceivability more accurately, catastrophic thinking makes it easier to see the threat of climate change clearly. ... The third

reason is while concern about climate change is growing – fortunately – complacency remains a much bigger political problem than fatalism. ... A fourth argument for embracing catastrophic thinking comes from history. Fear can mobilize, even change the world. When Rachel Carson published her landmark anti-pesticide polemic "Silent Spring," *Life* magazine said she had "overstated her case," and *The Saturday Evening Post* dismissed the book as 'alarmist.' But it almost single-handedly led to a nationwide ban on DDT.[16]

In light of the real possibility for the end of the world in the Anthropocene, the experience of a global catastrophe is intimately connected with how human beings understand and react to death. Death, fear, and the potential for annihilation, is a common theme in much contemporary political art of the Anthropocene. However, even though our very mortality is currently at stake, according to Theodor Adorno, we have lost a sensitivity to and an appreciation for death in contemporary society.

> The deterioration of the death metaphysics, whether into advertisements for heroic dying or to the triviality of purely restating the unmistakable fact that men must die – all this ideological mischief probably rests on the fact that human consciousness to this day is too weak to sustain the experience of death, perhaps even too weak for its conscious acceptance. ... The more our consciousness is extricated from animality and comes to strike us as solid and lasting in its forms, the more stubbornly will it resist anything that would cause it to doubt its own eternity.[17]

This has to change. The very real potential for death must be embraced and invited to the discussion. Literary critic and cultural theorist Sylvére Lotringer argued that a greater awareness and acceptance of death will be a key motivator in changing our attitudes about the Anthropocene: "We need to find a way to experience death in a different way. ... It would involve reconnecting with death instead of pushing it away. ... To think of an art of the Anthropocene without raising the question of collective extinction and death doesn't make sense."[18] Lotringer pointed out that at least for folks living in southern California, the sensory experience of earthquakes seem to do just that – make death by nature real and possible. The fires raging across the western states and around the world are also bringing home the actuality of death. Fear of dying through the violent agency of nature can play a key role in developing those insights. Perhaps the growing instability in weather patterns and the already dramatic intensity of weather events will result in a greater acceptance of the coming threats.

Philosopher Peter Sloterdijk also emphasized the role that fear of death can play in shaping cultural responses. He pointed out that several religions have long projected an End Times, when the world as we know it is predicted to end. These warnings of doomsday are currently mostly envisioned with passive acceptance, so the challenge will be to make the eschatology seem more real and immanent, the result of which may consequently lead to action

> Since effective narratives can only be organized in terms of their ending, the standpoint of the Anthropocene is identical to a powerfully moralistic narrative. In the narrative cultures of the West, this position was hitherto reserved exclusively for apocalyptic literature, which attempts to evaluate the world from the perspective of its end. … This is to say that everything suggests we ought to understand the terms 'Anthropocene' as an expression that only makes sense within an apocalyptic logical framework.[19]

Sloterdijk also argued for a new kind of proactive intelligence to address the Anthropocene – what he called "prognostic intelligence," an intelligence akin to an early warning system that is sensitive to the catastrophic, that

> asserts itself precisely in the gap between 'late' and 'too late.' … Whereas for a large part of human education to date people have had to 'learn from their mistakes,' the prognostic intelligence must become prudent before misfortune occurs. … According to Dupuy [*Pour un catastrophisme éclairé*], only experienced apocalyptics can perform reasonable future policy-making because only they are courageous enough to consider the worst as a real possibility.[20]

This is a prophetic intelligence that I believe is accessible to artists, whose task is to see ahead. Science fiction writers have been able to tap into this insight as they create new environments and explore human options within them. His call for this new intelligence is reminiscent of the Cassandra tale, only this time we must believe the warnings. The proverbial wolf is at the door.

Is stoking fears and dwelling on catastrophic potential a bad tool to use in raising awareness and motivating people to action? While some writers are concerned that fear may lead to paralysis and indifference, Canadian professor and filmmaker Michael Truscello noted that critics who promoted a less apocalyptic approach, "in their attempts to produce anti-catastrophistic messages they have understated the crises we face and promoted solutions that are dramatically inadequate, maybe even counterproductive."[21] In a similar perspective reminiscent of Kant, art historian Christine Guth suggested another use for fear – that by visualizing a fear of catastrophe, it could be disarmed. She noted that "Hokusai's print [of the great wave] has been especially helpful in grappling with representations of disaster in an interconnected world because by aestheticizing and exoticizing it also fulfills the function of distancing it."[22] Unfortunately, distancing can also lead to indifference.

However, Wallace-Wells concluded that there seemed to be little danger that stoking fears and dwelling on the catastrophic potential embedded in the Anthropocene would create panic or result in depression: "[T]here is no single way to best tell the story of climate change, no single rhetorical approach likely to work on a given audience, and none too dangerous to try. Any story that sticks is a good one. In 2018, scientists began embracing fear. ... *It is okay, finally, to freak out.*[23]

Research Fellow of the AMO Research Center Tomáš Jungwirth was also convinced that despite experiencing fear, fear could also motivate subsequent action resulting in a positive outcome: "Somewhat paradoxically, hope can be sought in the contagious character of the message of panic. ... Understanding the gravity of things is crushing yet, at the same time, inspires action. And so where all other motivations had failed to make us act, just perhaps, fear may succeed."[24]

Approaching catastrophe indirectly is an aesthetic technique philosopher Gene Ray explored. Ray turned to Adorno's *Negative Dialectics* to discuss Adorno's idea of a "negative presentation that evokes the catastrophe indirectly, obliquely, without pronouncing its name," forcing us to confront the terror with nothing to cushion us.[25] "Positive" images – say, of the piles of bodies in Auschwitz – do not "work" to move people to empathy or action, because they deflect critical reflection.

> Positive images give the impression of being complete, adequate, sufficient to the referent, whether that referent is a traumatic event or anything else. Negative presentations, by contrast, are honest about the elision; they at least show clearly that what needs to be shown cannot be, and lies beyond the presentation. And because they present themselves in this way, as ciphers that must be decoded, negative presentations invite and indeed demand a more active process of spectation – one involving close scrutiny and reflection.[26]

Ray concluded that negative representations like relics – suggestive images that might hint at a terrible trauma – are much more effective as signifiers. His example was an image of a blasted watch found in the rubble of Hiroshima. The image does not directly show us what happened, but opens up a narrative of possibility that will help us understand the deeper meaning. The oblique image, then, operates as an experience of formlessness that overwhelms any effort to make rational sense of a horrific event. It reaches through to our emotional affective core.

Research Fellow in the School of Ecosystem and Forest Sciences N. A. J. Taylor's reflection on the photograph taken by Associated Press photographer Richard Drew of a man falling out of the north tower of the World Trade Center on September 11, 2001, is also worthy of consideration in a discussion of death and the sublime. While this is a realistic image, its distance from the viewer and its lack of resolution of the trauma allow for an affective response. The predominant aesthetic symbol of the destruction of that day – the two towers crumbling – has been accepted and widely replicated without comment. Over and over the image has been burned into the cultural imagination. As a result, the monumental physical destruction of property shaped all ongoing narratives about the event. By focusing on buildings, the viewers avoided reflecting directly on the impact on human beings.

However, another, more human story can be told by Drew's very moving image which has been assiduously ignored. According to Taylor, immediately after the towers came down, "a small number of images sporadically appeared that betrayed the overarching narrative of September 11 as a world-historical moment,"

> Indeed, what binds a great number of these formerly orphaned images together are their preoccupation with the private moment - a man running for his own life, an office worker trapped, a fireman's expended body and mind. ... At the time "The Falling Man" was first shown in print and online news media, the public were so angered by its "exploitative" and "voyeuristic" depiction of one man's "private moment" that it was seldom seen again. ... [I]mages capturing "jumpers" plummeting from the towers were airbrushed out, and official statements employed vague wording to suggest that in fact no one had jumped, but rather had been blown out due to secondary explosions. In total, it is now estimated around 200 people jumped to their deaths as had, the man in that image.[27]

The fact that this image has been consciously removed from our consideration may be because death is primarily understood as a private experience in western culture. When it first appeared it generated a lot of anger, and this response may help us understand why it is so difficult to create an engagement with the potential for death that is implicit in the Anthropocene. Unfortunately, it is no longer just our own deaths to take into account, but the accelerating extinction of other species as well.

Death is an active presence in iconography of the climate crisis. The image of a dead baby kangaroo entangled in a barbed wire fence, victim of the fires in Australia in 2019, initiated widespread disagreement on social media. It was a shocking image, meant to induce empathy for the loss of wildlife and to inspire a sympathetic response to the global climate crisis. But if the responses induced horror and shock, was it effective in reaching people? Using another example, does the widespread sharing of images of abused animals result in an increase in activity to protect them? Did it result in increased giving to animal rescue and shelter programs? Or did these images serve to escalate viewing of violent or obscene material? People for the Ethical Treatment of Animals (PETA) has taken the position of condemning the sharing of abuse on the internets, encouraging people to report these images to authorities like the Federal Bureau of Investigation (FBI). They argue that they only serve to increase traffic at the offending sites which they believe is counterproductive. At the same time, PETA's site regularly posts stories and images of animal abuse.[28]

Recognition of human destruction of the environment and the subsequent grief over the resulting damage can also be a way of illustrating our place in the global warming hyperobject. Truscello's analysis of the art of Edward Burtynsky and Mitchell Epstein, for example, concluded that their images of massive ecological destruction such as mining waste, ruined landscapes, and polluted rivers reflect our new reality. Truscello referenced Morton's term "dark ecology" to describe how the images of polluted and ravaged landscapes can serve to bring home our role in creating the Anthropocene:

> Nowhere in these collections of photographs does one find an image that intimates a possible return to some form of pristine natural world; instead, viewers must confront the toxic future of oil refineries, hundreds of thousands of kilometres of pipelines, and other hyperobjects of petromodernity. … Morton's hyperobjects begin to articulate what I would call alien capitalism, an economic system whose materiality kills while dying, unleashes almost unimaginable toxicity even as its purpose or functionality wanes.[29]

We can approach the horror of these larger, collectively-caused damages through attentiveness to how we directly impact the environment at the local, the familiar, and the personal, levels. In this way I can better appreciate and accept the consequences of my role in the Anthropocene. My car exhaust is carried around the planet, soiling the skies; my microplastic pollution is destroying fisheries and bringing the deadly particles back into my body.

Artist Tamiko Thiel's installation, "Gardens of the Anthropocene," made a similar, but even more explicit connection between environmental pollution and our technological lifestyle choices, according to cultural and literary theorist Serenella Lovino. The surreal "Gardens of the Anthropocene," is a re-imagined garden "in which plants face a techno-biological mutation: while the "originals" are organisms capable of extracting nutrients from sunlight and soil, the mutant ones feed on the electromagnetic radiation of mobile devices and artificial structures such as road signs or street lights."[30] Thiel's installation captures the enmeshment of humans and the natural world. The garden, our human artifice, is planted with the horrors we have created, a true image of our current reality in the Anthropocene.

> This new "Anthropocene aesthetic" also has, therefore, a double face: on the one hand, we might become more sensitive to the transformations and threats we are exposed to, on the other hand we are almost "anesthetized" to the shapes they assume. These shapes might fall unnoticed or even acquire the status of a new sublime, just like the "colourful sunsets caused by particulate matter in the atmosphere."[31]
>
> "And so, if the Anthropocene is an "aesthetic event," art and culture must find a way to wake us up from the anesthesia and sharpen our perception of the predicaments that make our being-in-the-world. This might produce more costs now, but it might also be conducive to a world in which the awareness of the unjust will be in-built in the fabric of our biopolitical values, actions, and visions.[32]

Theil's work introduces an important concept for our consideration: the need to pay attention to the full life cycle of our technological creations. We ignore, for example, the plastic bag we throw "away" – forgetting that it lives on in its own manner and its own lifetime. Downstream everywhere there are impacts of our creations. We have responsibilities that will continue far into the future, according to Lovino: "That is one of the consequences of our becoming geological: all that happens, happens here and now; the ripples of our actions, as well as of our visions, will sooner or later reverberate right at our feet, directly in our gardens. … The end of externalities means that everything stays here: we have to deal with the consequences of what we do, of our actions as well as our visions."[33] The biggest challenge for artists, according to Jan Jagodzinski, "becomes how to address this event [the Anthropocene] that escapes easy visualization – like greenhouse emissions, for instance, that remain invisible, or natural processes that escape the radar of human consciousness."[34] Artists cannot just rely on images of the weather to convey the urgency of the global crisis. Weather is not global warming. Despite the increasing number of images on the nightly television news of worsening weather catastrophes – wildfires, hurricanes like Katrina and Maria, tornadoes and other cascading weather events – recent research noted that people accommodate themselves readily to heightened weather threats over time, opening the possibility that they will become indifferent to global warming and the awareness of changes over a long period of time.[35] I'm not necessarily convinced by this argument, however. There is a cumulative effect that can also take place. This occurred in American society during the Vietnam War. Demonstrations against the war were a direct result of seeing the daily death and destruction on the ground in Vietnam as American families sat around the dinner table watching the evening news on television. Images of the raging forest fires in Australia and California, in the Russian Taiga, and across Africa are driving home that these disastrous environmental events are likely linked to the ephemeral idea of global warming. Artists are faced with the challenge of representing the formlessness of an even bigger threat – grasping and articulating the immensity and uncertainty of the hyperobject. It is not outside of us, we are in this together with the planet.

The New Sublime
Deconstructing the Spectacle of Fossil Fuel

At the same time as artists are working to create an alternative aesthetic, we must continue to expose and critique the prevailing aesthetic of contemporary fossil fuel culture. There is a war of images going on before our eyes – a war that pits cheery images of life going on as usual against images warning of looming environmental disasters. Vested interests controlling the mainstream media and governmental messaging are working desperately to camouflage the reality of escalating dangers. There is a huge financial investment in continuing the denial which is driven by our continuing dependence on fossil fuel consumption.

The oil aesthetic is a capitalist aesthetic, and creating a new aesthetic will require dismantling the aesthetic that dominates our fossil fuel reality and the attendant consumer reality. Not until mid-twentieth century was a sophisticated critique of the driving force behind capitalist aesthetics developed by the artists and theorists of the French Situationist International (SI). Their critique exposed the political and aesthetic tools used by capitalists to fuel constant consumption through control of advertising images. In their analysis, they described the "spectacle" – an artificially constructed cultural world that stood outside of and was antithetical to individual human intervention and control.

The spectacle described the totality of the economic/social/cultural relationship that capitalism created to perpetuate imbalances of economic and political power. The spectacle was the constructed capitalist vision in which we now live, the bubble of friendly, entertaining oppression in which we move. There is general agreement that commodity consumption and the spectacle that has been created to perpetuate it control and continually reshape social values.

According to SI theorist Guy Debord, "The spectacle is not a collection of images, but a social relation among people mediated by images."[36] Echoing in some respects the early insights of Theodor Adorno and Walter Benjamin, Debord developed a sophisticated analysis of the construction of images in a technological society and their insidious control over all aspects of culture:

> At the technological level, when images chosen and constructed by someone else have everywhere become the individual's principle connection to the world he formerly observed for himself, it has certainly not been forgotten that these images can tolerate anything and everything; because within the same image all things can be juxtaposed without contradiction. The flow of images carries everything before it, and it is similarly someone else who controls at will this simplified summary of the sensible world; who decides where the flow will lead as well as the rhythm of what should be shown, like some perpetual, arbitrary surprise, leaving no time for reflection, and entirely independent of what the spectator might understand or think of it.[37]

The perceptions present in our environment – in this case the beliefs we share about oil – exert their control over us by shaping our collective behavior. As abstract representations disconnected from any social reality, these false realities are then fetishized and serve as pacifiers to divert our attention from the loss of control over our economic lives.

The SI argued that because the spectacle is all pervasive, it complicates the possibility for a clear and independent political response: "People are to a great extent accomplices of propaganda, of the reigning spectacle, because they cannot reject it without contesting the society as a whole."[38] In capitalist consumer society, spectators remain passive, eagerly anticipating a continually re-created spectacle that reinforces that passivity. The deconstruction of the spectacle, according to Debord, begins with understanding how our thinking is shaped by the spectacle, and how the petro-reality in the contemporary world is endangering humankind. Once the spectacle is apparent, "To actually destroy the society of the spectacle, people must set a practical force in action."[39]

The signs and symbols of the oil spectacle touch all aspects of our lives – psychologically, sociologically, spiritually, economically, and, of course, politically. They effectively shape and control us at all levels because they are designed to satisfy basic human needs such as the need for closure to experience, the need for social relationships and their representation of collective meaning, and the individual fascination with novelty. They also come to represent our culture as a whole giving us a sense of continuity across time with other people, a kind of false collective meaning. Countering this spectacle, making its manipulations transparent and deconstructing its meaning and power, therefore, is an important precursor to illuminating an alternative vision.

Artist James Koehnline's apocalyptic collage "Burning the Midnight Oil" effectively portrays the linkages between corporate power and threats to the environment that must be overcome. It combines ecological and political messages in images highlighting the interlocking relationships between fire and catastrophe, the power of fossil fuel in our society (represented by the United States Capitol building), and the entanglement of highways of the oil culture dominating the foreground.

We are immersed in the iconography of oil every day. We have become oil.

We live and breathe and eat and sleep cocooned in the petrol-reality. In a chapter entitled "the Aesthetics of Petroleum," in her book *Living Oil*, Stephanie LeMenager undertook an exploration of our passionate attachment to oil, and its alliance with happiness in American culture, and how we might go about defining an alternative aesthetic.[40]

She credits the peak oil movement with its message that oil will run out, with taking the first steps in illuminating the poisonous legacy of the oil culture. Building and implementing the alternative – living without oil – will be daunting: "[T]he resilient community must be flexible enough to reinvent its fundamental infrastructures, releasing itself from oil dependency to produce, largely by hand, all that it consumes."[41] Peak oil media brought into sharp focus an optimistic certainty that human imagination would figure a way for the society to continue moving forward without oil. "Philosopher Kate Soper has asked, more fundamentally, if there can be a 'new erotics of consumption or hedonist 'imaginary' that complements sustainability, in other words an affective intensity attached to limited growth, or no growth, that might rival, say, the embodied intensities of petromodern consumer culture."[42]

LeMenager asks us to consider four rhetorical questions about why oil is so bad – the answers contain clues to the aesthetics of fossil fuel addiction. If we can come to understand our enchantment with oil, and how we are hooked into the petro-reality, we can begin the process of unraveling those tentacles and breaking away to a new reality.

Fig. 2: James Koehnline, *Burning Midnight Oil Again*

1. *Why is oil so bad?* Because it has supported overlapping *media* environments to which there is no apparent 'outside' that might be materialized through imagination and affect as palpable hope.[43]

We are, she argues, captured by the spectacle and once we see our prison, we need to work to undo its power over us in order to free us to construct a new reality.

2. *Why is oil so bad?* Because of the mystified ecological unconscious of modern car culture, which allows for a persistent association of driving with being alive. ... The longing to really see the land, sharp desire to be more alive through its life, is an important twentieth-century environmental emotion connected to automobility.[44]

I've been watching the adverts on television for new cars the past couple of months. The ads always start with a single driver, driving fast on an open road that is usually winding off into the distance. There are no other cars on the road, and the scenery is unbelievably spectacular. Yes, the spectacle of oil is the car and our freedom to move through nature. The adverts continue with romantic music in the background, celebrating speed and movement – the merger of oil and fully living. In another ad produced for Marathon, the message linked gasoline consumption with patriotism – "Fueling the American Spirit." Several years ago, I was invited by a car company (along with other people), to review proposed technological updates for car interiors. We were each paid $100 to spend the day reviewing the latest hot button gadgets. At my exit interview I mentioned to the researcher that one big question was not asked: why are car interiors seeking to replicate the home environment, cocooned in music and visual stimulation? Comfort, internally focused. Everything to do except watch the road. It has only accelerated since then. Just recently, a car advert highlighted a full panel of screens across the entire dashboard – certain to distract the driver.

3. *Why is oil so bad?* Because its biophysical properties have caused it to be associated with the comic 'lower bodily stratum' ... In brief: Oil has been shit and sex, the essence of entertainment.[45]

LeMenager pointed to the iconic images in film in particular, that portray the striking of oil and the subsequent gusher as in the oil film *Tulsa*. Petroleum fires are spectacular, appealing to the attraction of fires generally. Sadly, even wildfires in California attract gawkers who line the roadways to watch the flames sweep up and down the land. Finally, she asks:

4. *Why is oil so bad?* We gather to watch.[46]

The spectacle of oil is, in the end, a collective experience, connecting us to one another, giving us a sense of belonging. In the United States most of us drive, and our days are filled with talk about our trips and our near accidents and the time we spend commuting, and how much gasoline costs, and the number of potholes in the road, and so much more. We don't share the same stories about taking the bus or walking. LeMenager concluded that we must transcend the world of oil and our immersion in its messages: "Decoupling human corporeal memory from the infrastructures that have sustained it may be the primary challenge for ecological narrative in the service of human species survival beyond the twenty-first century."[47] We must, in short, get out of our cars.

The transition away from fossil fuel consumption will not be an easy one. There may be violence and rage, there will be frustration, and there will be profound grief over the supposed loss of freedoms and an entire way of life. But we can already see that fossil fuels are killing our world. Fracking is polluting

Fig. 3 Asaf Hanuka, *Uncle Oil*

fresh water supplies, air pollution is still present but not as visible, and the hyperobject of CO2 buildup continues its relentless climb.

> The transition [away from fossil fuels] needs to be just as 'apocalyptic' as the many systemic breakdowns that are threatened by extreme human overrun and overreach. We are, and will be, confronted with things that will radically alter the world in ways we cannot control. We need to prepare a culture to respond to this challenge.[48]

One of the most difficult social and cultural challenges will be convincing the driving public to give up cars. The problem isn't just the car itself, it's the whole culture of the individual in charge of their possessions and their freedom to move about without consideration of anyone else. Fossil fuel culture is part of our "on demand," convenience culture. I will get in my car NOW, because I WANT to, and because I don't want to wait to walk ten minutes to a nearby bus stop. We are now being advised to buy electric cars – without any thought to the reality that electricity is produced by fossil fuels, and the likelihood of enough solar panels to fuel automobile movement is near zero. It also goes without saying that one of the seductive charms of electric cars – even if they are self-driving – is that the individual can travel in privacy. Of course, the roads will still have to be maintained.

Pathways to a Healthy World
Stories of Political Action

Art can move people to political action and taking action can mitigate the fear and paralysis that theorists suggest is inhibiting a meaningful response to the climate crisis. Art is, after all, a making and a doing itself; an action and an important way of experiencing ourselves as actors, as creators, in changing the relationship between human beings and the planet.

In *The Politics of Aesthetics*, French philosopher Jacques Rancière drew attention to the similarities between politics and art: "Politics and art, like forms of knowledge, construct 'fictions,' that is to say *material* rearrangements of signs and images, relationships between what is seen and what is said, between what is done and what can be done."[49] Rancière argued for the marriage of politics to art, of art to action – the integration of politics and a sublime experience, creating a rupture with one world to open a way to another:

> The dream of a suitable political work of art is in fact the dream of disrupting the relationship between the visible, the sayable, and the thinkable without having to use the terms of a message

as a vehicle. It is the dream of an art that would transmit meanings in the form of a rupture with the very logic of meaningful situations. As a matter of fact, political art cannot work in the simple form of a meaningful spectacle that would lead to an 'awareness' of the state of the world. Suitable political art would ensure, at one and the same time, the production of a double effect; the readability of a political signification and a sensible or perceptual shock.[50]

Storytelling is a very effective way of communicating the complex visions and messages of political change, and indeed, can shape activism. According to Boris Sax,

> The toughness and flexibility of storytelling make narrative especially important in periods of transition, particularly at the origin of new religions, ideologies, institutions, or technologies. Storytelling is especially effective because it has a sensuality that places it especially close to experience. Tales evoke sights, smells and sounds, while philosophies and precepts usually do not. For another thing, storytelling has great flexibility. Narratives can more easily be adjusted to different eras and circumstances than can rules or ideologies.[51]

Art of the Anthropocene calls for a politically-oriented narrative on behalf of the environment. Stories, as Jennifer Rae noted, have often moved people to take political action. "The affective capacity of art has been valued for centuries to inspire and motivate people into response and action. Historically, a wide-range of artists and artworks have helped to catalyse social change either intentionally or as a result of their work."[52] Rae pointed to Charles Dickens' novels, Harriet Beecher Stowe's *Uncle Tom's Cabin*, Pablo Picasso's painting *Guernica*, Randy Newman's song "Burn on Big River Burn On," as examples of art that have motivated political action. Pete Seeger's efforts to clean up the PCB, mercury and sewage pollution in New York's Hudson River in the 1970s were inspired by his song, "My Dirty Stream." His messages led directly to the passage of the federal Clean Water Act of 1972.

According to cultural theorist Stevphen Shukaitis, aesthetics should be grounded in developing human [and nature] relationships and a political commitment to action, rather than passive reflection.

> Bourriaud argues that today the designation of art seems to be little more than a "semantic leftover" which should be replaced by a definition like "art is an activity consisting in

producing relationships with the world with the help of signs, forms, actions, and objects" (Bourriaud, 2002: 107). From this Bourriaud tries to recast the critical function of aesthetic intervention, arguing that rather than being based on forming imaginary and utopian realities, artistic intervention is aimed at forming living models of action and being within the existing world.[53]

Artists also have a responsibility to educate the general public, perhaps one of the most important political activities that they could undertake. The context in which art of the Anthropocene is presented becomes critical. Where can people best be reached? Traditional venues like museums is one option. Other venues might include classrooms, murals in public buildings, posters, or social media platforms. Accessibility to information from a well-respected source can move people to action.

I recently had an occasion to visit the new Bell Museum of Natural History on the campus of the University of Minnesota. It is the only natural history museum in the state. It is here that thousands of visitors come every year to learn how the earth was formed and how it changed over time – a perfect venue for communicating about the Anthropocene. The physical layout of the museum leads the visitor through a series of geologic time spans where they are confronted first with the starry expanses of the cosmos, then the formation of the landforms, moving through the lower life forms, to a giant mastodon and an equally impressive prehistoric giant beaver. The tour continued with dioramas portraying contemporary life forms indigenous to our state's ecosystems.

I looked around at this point for an exhibit on the Anthropocene. Alas! Global warming was tucked away in a corner (out the flow of the museum's traffic) with no signage directing visitors, or even that the corner was part of the larger geological timeline exhibit. A conversation with the director of the museum revealed that the topic was not covered because it was politically sensitive, and the museum did not want visitors to have an "upsetting experience." I was dismayed, to say the least, that visitors and especially young people and children were not learning about the climate crisis. It seems to me that we have an obligation to give people "bad news" – particularly since foreknowledge could lead to taking action in the present rather than postponing it for some future time. In the United States we already have many public settings where painful topics are covered in a sensitive way: the Vietnam Memorial, the Holocaust Museum, the National Museum of the American Indian, the National Museum of African American History and Culture with its narrative of slavery, or even the many battlefields of the American Civil War scattered around the country. One would think that addressing such an urgent topic as global warming in a scientific setting like the Bell Museum would be an imperative.

On a more positive note, the newly founded Climate Museum in New York City is committed not only to confronting the challenges of the Anthropocene head-on, but taking a political message to fellow New Yorkers. For example, in 2018 the museum commissioned artist Justin Brice Guariglia to install a public art exhibit in several local neighborhoods. The museum described the "Climate Signals" project as follows: "The show consisted of ten solar-powered highway traffic signs installed in parks and public spaces, sparking dialogue and drawing passers-by into the climate conversation. *Climate Signals* promoted understanding of the local impacts of climate change, its intersection with other urban challenges, and the importance of climate action by cities." In an in-depth article exploring the impact of the project, Emily Raboteau concluded: "We need to get more human-focused but not more anthropocentric. As a human, I'm on the same ontological playing field as garbage and galaxies – not above. As a visual artist, I'm trying to engage with a huge thing operating on several levels requiring several languages. This is an exceptionally urgent problem. It needs to get out into the broad public and raise consciousness. That's the responsibility of artists and writers. Not the corporations."[54]

We are Home
Context, Place, and Belonging

> A good second step is to take ourselves away from the domain of philosophy into the realm of direct encounter, to experience the presence of non-human activity and purpose. As Patrice Jones recommends, we should focus our awareness on the active natures of our non-human kin and marvel at their ingenuity.[55]

Describing the natural world and our place in it is a complex undertaking. How can place be portrayed? Is place where my heart resides, or is it an ecological region that does not follow national boundaries? Can it be the apartment building or the neighborhood I live in? Or is my place my backyard, my garden? We all live in and connect with many places. I live in Minnesota, a state that includes three separate ecosystems and borders two great bodies of water – Lake Superior and the Mississippi River. My home is also the middle of the North American continent – the Midwest, too, can be my place, where fields of wheat and soybeans and corn stretch as far as the eye can see and where the river connects me with all the states between here and Gulf of Mexico.

How far can real connection to a place extend? Imagining the ocean as a place to care for could be difficult for many of us – especially if we live far from the shore, but now we have to care because the radiation fallout from Fukushima could appear in the salmon on my dinner plate. And what about our air? Is this my place, too? Should I care about a wall proposed for the border between the United States and Mexico that will divide habitats for animals and plants and interfere with their migrations?

Place is not just how I feel about where I live, it is part of its own environmental identity, it is the shared environment about which we can speak truly and fight for its needs. The climate crisis demands that we clearly stake a claim in our place and commit to defending it, even if it is a place where the waters in the lake are polluted with herbicides and pesticides, or where we see the sea levels inexorably rising, or the hillsides burning in the drought season. Place speaks for itself, for what is actually there, and my place is what I am challenged to experience and interact with in a meaningful way. Lawrence Buell has explored the sense of place as an ongoing discovery process:

> Altogether, it seems that place-consciousness in literature, and most especially the consciousness of the nonhuman environment as a network enfolding human inhabitants, ought to be considered a utopian project that realizes itself, in its more instructive forms, not as a fait accompli but as an incompletion undertaken in awareness that place is something we are always in the process of finding, and always perforce creating in some degree as we find it.[56]

How do we describe the natural world? Sometimes we romanticize it. We love to sit on the beach and enjoy a sunset or image living in Thoreau's cabin on Walden Pond. Some of us assign the natural world divinity status, calling it Gaia or Mother Earth. Others adopt more scientific or systemic descriptors such as the biosphere, or the web of life to describe its complexity. Artists have often personified nature, assigning human attributes to its beings. The concept of Gaia is a good example. Mother Earth is portrayed as having the human qualities of nurturance, of abundance. But is it even our place to assign a name, or a function, or a description? Does doing so disrespect or ignore the integrity of the other, or does it limit how we interact? Other beings may have their own agendas, and may be totally indifferent, or even hostile, to human interests and the ascription of anthropocentric characteristics. Naming also fixes the environment in the parameters of what we think the term means. These descriptions are often static, overlooking the fact that cascading interactions – not fixed characteristics – are the hallmark of the hyperobject. For every action, there is potentially an unknowable and unpredictable ecological reverberation, reflecting a changing and dynamic nature.

Living in a place with other beings must be grounded in a deep awareness and respect for the agency of the others. This will allow us to appreciate the synergy operating to maintain the equilibrium of the environment. We must become attuned to the intricacies balancing the competing agencies. The agency of nature – including our own agency as a part of nature – must not be a matter of competition for survival, but life forces working together to accommodate themselves to the others surrounding them. This is the intelligence of the whole. Timothy Morton reminded us of our obligations to the others in our shared world: "The aesthetic experience is about *solidarity* with what is given. It's a solidarity, a feeling of alreadiness, for no reason in particular, with no agenda in particular – like evolution, like the biosphere. ... Ecologically explicit art is simply art that brings this solidarity with the nonhuman to the foreground."[57]

Seeing the other beings with whom we live as partners requires a relationship that is reciprocal and respectful and grounded in a place. From deep connections to our shared place will come a more confident responsiveness and hopefully, action to protect all the residents of that special ecosystem. Learning and appreciating and acting on this knowledge is a matter of life and death. Finally, seeing yourself as an actor in a story is as important as being engaged with the other beings. As part of the story you will be part of creating a new world, a new future out of the chaos of the collapsing world.

Artists can minimize the desire to personify and objectify nature by immersing themselves in the real and actual ecosystems surrounding them. The appreciation of our complex interconnectedness is not just a call for some idealized Kumbaya moment with the natural world. Arnold Berleant invites us to look more deeply into our relationship with the environment: "engagement is central to all aesthetic appreciation. ... Aesthetic engagement leads to appreciation that is not contemplative, not subjective, and not an exclusively mental act but an activity that requires bodily participation."[58] Such attentiveness to the richness of an environmental context allows for listening to the others and embeds their stories in real experiences. Finding and exploring local place can be the beginning of developing a new relationship with the natural world. But narratives must not just portray nature as background, or grounding, an indifferent backdrop or setting. Nature is a vital, constantly changing being, requiring engagement.

The hyperobject of global warming is one in which we all reside and that surrounds us, envelops us and is a special kind of "place." How human beings understand our role there is problematic, to say the least. It is not a place outside of us, and while it may seem contradictory, grounding in the real world is critical to understanding the unpredictable nature of global warming. Grappling with portraying its immensity will be one of the most important challenges for artists because it is really materially unknowable. We have only its effects to witness across time and in the abstract. Connecting us with the complexity of multitudes of agents of the planet like soil and air and water, will be a formidable challenge. As Jan Jagodzinski observed, "[T]he challenge becomes how to address this event [the Anthropocene] that escapes easy visualization – like greenhouse emissions, for instance, that remain invisible, or natural processes that escape the radar of human consciousness."[59] As my grandson, Robert Yount observed in a recent conversation, "This issue is hard in multiple ways; it's not just that we can't see the impact of our individual actions, or that we can't see global warming on an individual basis. We also won't be able to see our own individual impact of what we do differently."[60]

We have to learn to see in a new way. Psychologist Robert Jay Lifton explained the shift in awareness that will be required as we transition from the Enlightenment subject/object way of understanding reality, to a more integrative and complex reality. To understand the immensity of the hyperobject will require an integrative skill, the ability to see the "big picture." He described the difference as the distinction between "fragmentary awareness" and "formed awareness." According to Lifton, in an interview with environmental journalist Diane Toomey,

> Fragmentary awareness consists of a series of images that may be fleeting. ... Formed awareness is more structured awareness, so that there's a narrative. There's a cause and effect. ... A narrative, a story. And there's a parallel with climate. With climate images, when they're fragmentary, we may have an image of a storm here, of sea rise here, a little bit of flooding there, the drought. But when that becomes a formed image involving global warming and climate change, we take in the idea of carbon emissions leading to human effects on climate change and endangering us. And in that same narrative, there can be mitigating actions to limit climate change.[61]

Narratives of the new sublime can create connectedness and shared meaning by pointing out the complex elements in our environments and introduce them in digestible pieces in real time. At the same time, context also requires a connection to a particular place, and the most effective stories are those based on real experiences. As biologist Carl McDaniel observed, "A culture creates its present and therefore its future through the stories its people tell, the stories they believe, and the stories that underlie their actions. The more consistent a culture's core stories are with biological and physical reality, the more likely its people are to live in a way compatible with ecological rules and thereby persist."[62] We cannot respond to an abstract world or an abstract hyperobject. Global warming is real when we can see the traces of its effect in our neighborhood, in our region's production of food, how clean the waters are that might have to provide our drinking water, or what mining waste is precariously stored up-river. Sustainability design expert Bill Reed also highlighted the importance of the "story," and the necessity that it be embedded in place:

> First, history has shown that we will not sustain the will needed to make and maintain the needed changes, day after day, without evoking the spirit of caring that comes from a deep connection to place. A clear cultural narrative is needed to convey the connection to a particular place. Second, discovering the story of a place enables us to understand how living systems work in a particular place, and enables us to bring greater intelligence to how humans can then align themselves with that way of working to the benefit of both. Third, the story of place provides an integrative context that helps maintain the spirit and vitality of holding a collective and meaningful purpose. Finally, the story of place provides a framework for an ongoing learning process that enables humans to co-evolve with their environment.[63]

As the climate crisis deepens, local political and economic activities will become more important. The forces currently in place have a vested interest in maintaining the status quo and will go to great lengths to minimize any threats to their profitability. Between 2017 and 2019, for example, the United States federal government systematically suppressed scientific data and rolled back over seventy environmental regulations. Some states, however, immediately initiated their own mitigation efforts. Local knowledge of damages to local environments will allow for more immediate political action.

In 1969 Santa Barbara California photographer Dick Smith covered the Santa Barbara oil spill – a seminal moment in modern environmental disaster events. According to Stephanie LeMenager, who reviewed his photographic archives for her book, *Living Oil: Petroleum Culture in the American Century,* Smith was deeply moved by his photographs witnessing the deaths of the birds and animals – who mostly died from suffocation as a result of being covered in the oil. Smith was especially dismayed by how the government covered up the deaths of mammals, leaving the public unaware.[64] Local witnessing is crucial, LeMenager believed, to bring home the realities of the climate crisis: "Without looking at animals dying in oil, … thinking about such victims, and about, say, the regulation of off-shore drilling, [response and action] maybe also be less likely."[65] The damages must be made manifest and shared with the larger public, and they must be presented in a forceful manner.

The Power of Affect

Affect – the experience of feelings and emotions that shape our behaviors – plays a central role in the development of every narrative. A visceral commitment is necessary for sustaining permanent change and action. What connects the viewer and a work of art, for example, are the resonances of the aura of the work, the bringing together of viewer and object into a new experience. This experience is neither rational, nor emotional – it is an affective response. Narratives, of course, provide a good example of how to engage the affects; we are "caught up" in a story, following it to the end.

Affects can be manipulated such that we are moved to positive or negative responses. Fascism, for example, attracts people into an affect-driven culture of fear and authoritarian safety. Advertising preys on our innermost desires. Are there potential dangers in an over reliance on appeals to powerful feelings? Cultural critics have warned us for over a century (Adorno, Horkheimer, Jameson, among others) that we are living in a culture of affective overload. We are bombarded on all sides, our feelings pulled first one way then the other. Media-saturated cultures are characterized by continual appeals to our affects – advertisements, spectacular entertainment, news that is always "breaking" and thus demanding an immediate response. The result is over-stimulation, numbness, distraction, shortened attention spans, and finally, inaction, exhaustion, and indifference leading to inaction. Cultural theorist Clair Colebrook, in "Earth Felt the Wound: The Affect Divide," has suggested that the impact of overstimulation on responses to the environmental crisis has resulted in a general paralysis of affect response and the failure to engage in subsequent activism.[66] As knowledge of the negative effects of the Anthropocene increase,

> the more disengaged the intensity appears to be. 'We' late near extinction humans appear to be addicted to witnessing annihilation, to the feeling of near-death or post-human existence, and yet have no intensity: it does not prompt us either to action or to any sense of what a post-human world would be. On the contrary, the more evidence, imagery, feeling and "experience" of a world without humans is displayed, the less affect or intensity occurs.[67]

Philosopher Janae Scholtz agreed with Colebrook's conclusions:

> What also becomes clear is that affect in and of itself is no panacea – it isn't the case that affects can 'save' us from an over-intellectualized, over rationalized world, or that they will necessarily be agents of change in our perceptions or behaviors, because affect has already become the mode of exchange in our current

economy. In fact, the problem is much deeper – the oversaturation of affect actually means that we have become impervious to its effects.[68]

Facing the existential threat of the climate crisis, however, may override the ennui and boredom of this incessant, late capitalist entertainment and information programming and motivate us to action. A recent study of social media sharing behavior by psychologist William J. Brady and others, suggested that appeals to moral and emotional language can break through the mind-numbing affective clutter and trigger a response that motivates people to communicate their deep concerns for political issues in particular.

> In three studies using tightly controlled lab experiments with increasing ecological validity and linking these data to real Twitter communications, we found that (a) moral and emotional language both capture attention to a greater extent than neutral language, and (b) such attentional capture potential in words is associated with real-world patterns of retweeting on Twitter. These data shed light on the cognitive underpinnings of the spread of moralized content online, which can help explain how political leaders, disinformation profiteers, marketers, and online activist organizations can spread content by capitalizing on natural tendencies of our perceptual systems.[69]

As I argued in Chapter One, an appeal to affect is an essential part of every artistic expression. Human beings are more likely to react in a non-rational manner, so rational and logical arguments, or the weight of scientific consensus will not likely move people to save the environment. The Critical Art Ensemble argued instead, that "Argument has to be augmented with strategies and tactics that mimic the insights into nonrationality of behavioral economics, marketing, and advertising, in order to relate 'small-*t*' truth through means other than reason, and ultimately to bring the participant back to a position of reason."[70] The authors shared several examples where affective narratives have had a powerful impact on environmental thinking: Rachel Carson's DDT story of how the poison affects people; Paul Ehrlich's labeling overpopulation as the "population bomb"; the shock in response to the discovery of the hole in the ozone layer; and the discovery of the Great Pacific Garbage Patch that first alerted humans to trashing of the oceans.[71]

The deadening of affect can be approached differently, challenging artists to create new connections and resonances. Advertisers refer to it as transcending "the clutter" of competing media. Janae Sholtz pointed to John Cage's use of silence in his music to interrupt the "noise" of contemporary culture to create a different approach to affect. "[W]e can engage a new potential – a space in which humanity can become that which understands itself from a new conception of immanence and affect, to become a people sensitive to open, dynamic systems of intensities, forces and multiplicities. ... to think about the human, or being human, differently, as an open possibility constantly bombarded by and in tandem with myriad of forces and affective relations to other beings, human and otherwise."[72]

Timothy Morton also highlighted the important role of affect in art in the Anthropocene. Reaching out with our feelings and emotions will break through the separation between self and nature:

> The trouble with the PR approach, or the reason-only approach (its twin in some ways), is that human beings are currently in the denial phase of grief regarding their role in the Anthropocene. It's too much to take in at once. ... [H]ow do we talk to the unconscious? Reasoning on and on is a symptom of how people are still not ready to go through an affective experience that would existentially and politically bind them to hyperobjects, to care for them. We need art that does not make people think (we have quite enough environmental art that does that), but rather that walks them through an inner space that is hard to traverse.[73]

Restoration, Regeneration, and Hope
Listening to the Planet

We can no longer wait in the shadow of a pending catastrophe.

Despite the fact that the Anthropocene looms large in the rear view mirror of our lives, with our new awareness we can begin to imagine – and then create in collaboration with our brother and sister beings – a sustainable future. Artist James Koehnline calls us to this important task, using the tremendous power of the imagination to break through the old, discarded stories and unravel the spectacle to change the world.

> The fairy tale backdrop on the stage of our attention is full of tears and full of holes and is whipped by a black wind howling obscenities in an endless, starless night, and we cry out: On with the show. Tell us new stories. Patch and mend as we go, but keep the stories coming for the duration of our allotted span. We cannot live without our stories, and we'd rather not die without them. But they needn't all be prick tales and sky

god projections. We have a bigger bag of tricks than only that. We can tell better stories than those. Can't we?

This is just a variation on an old theme. The image of the torn fairy tale I owe to Kazantzakis in his play, "Buddha" which I read 35 years ago. The magician in the tale says,

> "I've come. … Man must have need of me again. The fairy tale that conceals the truth must be torn and hanging like a rag again. Man saw what was hiding behind it and became frightened. He cried out my name to the wind three times: Help! Help! Help! Help is my name. And here I am, with my unconquered army - the multi-colored, multi-winged, multi-eyed regiments of the imagination that fly over the dung heap called man's mind. My army will destroy the breakwaters of the brain. It will disrupt and pillage all certainties."[74]

Artists can take up the challenges before us and communicate the pathway forward, but first we have to believe that alternative pathways and responses are possible. In taking action lies the potential for a hopeful outcome; in paralysis lies certain failure and even death – to ourselves, to the world around us.

How can we address the future without despairing? Initially, denial and depression are common responses to the realities of the Anthropocene. It's time to move to the next phase – taking personal responsibility. Hope lies in making a commitment to the regeneration of the planet – at least to the extent that we can. We are in the midst of a crisis, and happiness seems only a distant possibility. The hope – and the joy, and the happiness – will come when we take up the task of restoration to bring the environment back into balance.

We also have to be realistic as we move forward. In 2019, environmental research scientist Jennifer Marlon and others, completed an interesting study of how hope and doubt affected people's responses to the climate crisis. The study drew a distinction between "constructive hope" – that people were becoming more aware of the need to respond to the crisis and were ready to take action, and "false hope" – the belief that a god or natural processes would solve the problem. Constructive hope generated more awareness and political activism, they argued, serving to "motivate effort, goal achievement, and adaptive responses in the face of adversity." [75] False hope, on the other hand, led to skepticism and indifference.

The authors emphasized that communicating a hopeful attitude was important: "Such stories would focus on seeing others taking action, information about changing social norms and growing awareness among the public, information about the co-benefits of reducing global warming (e.g., clean air, economic growth, technological advancement) and stories about local to global initiatives that are succeeding."[76] We won't be able to fix everything, but we can make changes on individual and collective levels both, to advocate for new economic and political structures that recognize that nature now has a seat at the table.

The restorative process will require that we move beyond our fears, our despair, and our denial to pick up the pieces and move forward. Jem Bendell criticized writers who did not challenge the denial narrative because they believed that people will be unable to cope with the fear that warnings of an impending catastrophe will bring. He argued instead, "There is some evidence from social psychology to suggest that by focusing on impacts now, it makes climate change more proximate, which increases support for mitigation."[77] According to Bendell, we should not avoid confronting fear and despair, for out of that despair a new hope may emerge: "Confronting our fears and our hopes, can be very empowering: In my work with mature students, I have found that inviting them to consider collapse as inevitable, catastrophe as probable and extinction as possible, has not led to apathy or depression. Instead … something positive happens. I have witnessed a shedding of concern for conforming to the status quo, and a new creativity about what to focus on going forward."[78]

How will I proceed? I will begin by listening to the planet, hearing its cries of distress and begin to act in a collaborative manner with the planet that sustains me.

What can you do? What can we do together?

Fig. 4 Angel Boligán, *Our Flora and Fauna*

DO YOU SEE WHAT I SEA?
THE SEA IN POLITICAL ART OF THE ANTHROPOCENE

"The sea has a story to tell. It is a story that
it has been telling for thousands of years. It
can tell it in many different ways."
– Brett Bloom[1]

VISUAL IMAGES HAVE A POWERFUL IMPACT IN ALL CULTURES, AND THEY PLAY AN increasingly significant role in shaping the contemporary political reality in which we all live. The phenomenon of the internets and the sharing environment of social media have spread political images worldwide. In an environment where images are used profusely to convey complex political ideas, it becomes even more important to understand the nature of imagery, the meaning conveyed through images, their collective impact on our lives, and how we might play a greater role in interpreting the images that surround us.

To make sense of the visual environment, human beings collectively create and share symbolic constructions of meaning. Icons are dynamic representations, designed for universal recognition, contemplation, and the sharing of recognized collective meaning; memes are active languages designed for social exchanges and viral transmission. These representations also have a powerful political function – often inspiring people to think and act in specific ways. Packed into these representations, via the creation of visual narratives, are layers of meaning, feelings, affects, hopes, and dreams. In political art, there is another crucial objective – inspiring people to action.

How can images of the sea help us see and experience the Anthropocene? This chapter focuses on contemporary images of the sea portrayed in political drawings, cartoons, and posters – specifically, representations of the *agency* of the sea as an awe-filled active force in a threatening environment. The sea is, I believe, a keystone symbol that has the potential to help us experience the power of the Anthropocene as a hyperobject. Timothy Morton's hyperobject is a phenomenon that is difficult to visualize in its immensity. It is not easily comprehended and remains a mystery, the awesomeness of the unknown. I am interested in particular in how the sea might be "seen" as an actor in the human response to the Anthropocene. The sea lies deep in the human subconscious, a

Fig 5: Leonardo da Vinci, *The Deluge*, 1517-1518

remnant of ancient terrors of immense powers. The very foundation of aesthetics revitalizes these ancient forces in contemporary representations that make manifest the agency of the natural world.

Artists have long been fascinated with portraying the dark side of the sublime and in particular its connections to the sea. For example, toward the end of his life, Leonardo Da Vinci created a series of eleven apocalyptic drawings on the theme of the deluge. This series is "among the most enigmatic and visionary works of the entire Renaissance. Modest in size and densely worked in black chalk, they cannot be arranged as a single coherent sequence, but each shows a landscape overwhelmed by a vast tempest," according to the Royal Collection Trust.[2] The Minnesota Marine Art Museum in Winona, Minnesota exhibits a stunning variety of artworks focusing on marine life and water. Artists in the collection include: Turner, Monet, Renoir, Van Gogh, Cassatt, Gauguin, Picasso, Matisse, Kandinsky, O'Keeffe, Homer, Andre and Jamie Wyeth, and many more.

One of the most prolific artists of sublime seascapes was the early nineteenth century English painter J.M.W. [Joseph Mallord Willliam] Turner. His powerful images of shipwrecks, violent storms, and disasters conveyed the fury of the sea. His style – realism bordering on abstraction – embodied the essence of human fear of the sea in paintings like "Snow Storm - Steam-Boat off a Harbour's Mouth," "The Storm," and "The Slave Ship." Ivan Aivazovsky, a

Fig. 6: J.M.W. Turner, *Shipwreck of the Minotaur*, 1810s British Museum

Fig. 7: Ivan Aivazovsky, *The Ninth Wave*, 1850, Russia State Museum, Saint Petersburg

mid-nineteenth century Russian-Armenian painter is not as well known, but he, too, produced a large body of works, over half of which focused on representing the agency of the sea with titles such as "The Ninth Wave," "Hurricane on a Sea," and "Shipwreck."

Sublime images of the sea re-ignite human fears about the uncontrolled powers of nature. There is even a formal term describing this fear of the sea – thalassophobia. The sea is one of the most significant and persistent environmental icons – especially in its turbulent, fearful aspect. Symbols of drought and fire do not seem to resonate with a similar affective power.

Our impressions of the sea have changed over millennia – from moving watery representations of anger, fear, and awe, or a sea filled with monsters; to a highway of global expansion; to serene seascapes in later centuries. But the fear of the sea persisted as an underlying theme. While the sea itself was the initial symbol of this fear, lurking under the sea were all manner of unknown beasts threatening humanity [Figures 8-11]. Early maps, for example, portrayed fantastical images of mythological serpentine beasts populating the waves. From time immemorial, fear of the sea was endemic to those cultures where people lived close to the sea. For people living on the shores, the threats were very real. In medieval Japan, for example, permanent markers were placed along the shore near the Fukushima nuclear power reactor, to warn people of previous tsunamis, and to avoid building there, reflecting a deep respect for the latent

Figs. 8 (above) and 9 (right)
Detail, Medieval Maps

Fig. 10: Olaus Magnus, 1555

Fig. 11: Hans Egede, 1734

power of the waves. The universal fear of the sea meant that human settlements worldwide, up until the sixteenth century and beyond, were established far from the shore, away from any real or perceived threat.[3]

The active power of the sea, its agency, and whether there was malign intent driving that agency, was the source of that fear. Its power was beyond human control, the living presence of the potential for disaster. The life-threatening force of waves both against the shore and ships at sea was deep in human consciousness.

There are many terms in the English language to describe powerful weather systems: blizzard, hurricane, tornado, typhoon, flood, cyclone, tsunami. Several relatively new terms have been introduced to this category, perhaps to reflect the experience of mega weather phenomena – "atmospheric river," "flood drought," and "bomb cyclone." "Atmosphere river" refers to weather patterns over the Pacific Ocean dumping rain on California – they used to call it the "Pineapple Express," but atmospheric river better describes the onset of torrential rainfall that brings catastrophic flooding. "Bomb cyclone" refers to a rapid drop in low pressure, specific to a geographic area, resulting in violent storms. "Flood drought" describes the phenomenon of prolonged drought, followed by catastrophic flash flooding, such as seen in desert areas. None of these land-based phenomena, except perhaps the tornado, has visual symbols instilling fear and terror in the face of the power of nature – it is primarily the sea that inspires sublime images of terror.

Iconography and how cultures imagined the sea changed, beginning in what European countries came to call The Age of Exploration. Fear of the sea was replaced with indifference and the confidence that new sailing technologies had vanquished the dangers. Soon sailors had the power of the steam engine to assert human dominance of the waves, and to realize their imperialist expansion objectives. The oceans became the highways of military and economic conquest. New, imperial cities were established on the shores close to the sea, from which the invaders could defend their conquests from incoming competition, replacing the cities that had been built further upriver in safe locations away from the ocean.[4] According to anthropologist Jake Phelan, by the eighteenth century, "In general … [the sea was] to be traversed not described. The sublime sea of the late eighteenth-century was the first time the sea was seen as more than a stage. … There followed the Romanticism of the nineteenth century, the glorification of untamed wilderness and man's dominance over nature."[5] Iconography celebrated vast and beautiful seascapes, idyllic beaches, and sunsets reflected over calm waters.

In mid-twentieth century in the United States, the sea once again became the source of cultural anxiety and fear. According to English professor Nicholas Jenkins, the "avant-gardes and mass culture were, to a degree which has not yet been understood, obsessed by the sea."[6] Further, "A sense of fascination with depth, width, and vastness had become an important visual experience with strong moral dimensions for American audiences as they looked at films and paintings in the wartime and postwar period."[7]

Jenkins proposed two different views of the sea to explain the changing perceptions over the course of the twentieth century: a modernist period (1910-1930), where "the sea is … confidently linked with heroic or utopian moods," and later, "in the mid-century period the 'oceanic feeling' usually had to do with terror, desolation, anguish. … [T]he American ocean became a place where men – and it *is* overwhelmingly men in this period – went to kill and die and never merely to cross."[8] The popularity in the 1950s of the American television program "Victory at Sea," set against the background of the naval battles of the Second World War, served to reinforce the image of the sea as the site of military conquest and western culture manifest destiny. The victory narratives dramatized the supremacy of America in naval wars across the world's oceans: the Pacific, the Atlantic, the Mediterranean, the Indian, and the Black Seas. Fear of the sea lingered, however, and was an important component of mid-century perceptions of the sea linking the sea to death: "no one controls the sea. It can destroy but never to plan. It is lethal, but no one can make it a weapon. Its unpredictable violence is also undifferentiating and unlimited. Nobody made it."[9]

Mid-twentieth century fascination with the sea was also reflected in an era of expanding underwater exploration, the oceanic trip of the *Kon Tiki* which traced a hypothetical trans-Pacific human diaspora, and the publication of Rachel Carson's, *The Sea around Us*. Written for a lay audience, her book introduced the public to the vast richness and diversity of ocean worlds. It was also an early alert of how the United States (and eventually the rest of the world) viewed the ocean as a dumping ground. Carson reported, for example, that the Atomic Energy Commission had dumped "low level nuclear waste" in metal drums off the eastern coast.

By the late twentieth century, the sea was again viewed primarily as surface – its purpose was to convey the burgeoning movement of goods highlighting the globalization of capitalism. Photographer Alan Sekula captured this capitalist conquest of the sea in his exhibit and catalogue, *Fish Story*. What is striking about his images is the relative absence of the sea itself. It is portrayed only as backdrop, as vehicle to chronicle the story of the global movement of container ships, consumer goods, and capital.[10]

Early in the twenty-first century, the ancient fears of the sea were re-ignited in a visual response to the awakening awareness and fear of global warming and the unknown consequences of impending climate crisis. The sea once again was assigned metaphorical terror. Rapid decline of fish and sea mammal populations due to human predation, the global destruction of coral communities, and the appearance of massive floating fields of garbage in the middle of the Pacific Ocean raised alarms worldwide. Humanity had come full circle: from respect embedded in fear, to containment, to romanticization, to imperialistic triumph, to benign representation, and finally, to a return to the ancient terrors of fearsome power.

Images of Thalassophobia (Fear of the Sea) in Political Art

If artists want to raise awareness about the climate crisis, what is the role of political art in shaping perception and action? An examination of contemporary images of the sea in political art could provide new insights into how artists are using images of the fear of the sea to address the human agency of the Anthropocene. My research into images of the sea shared on social media led me to identify several powerful memes – the most common of which were related to the fear of the sea.

One of the strongest human impulses is the "fight or flight" response to fear. Fear is a powerful motivator of action, triggering an awareness of danger or threat, and motivating us to action. If political images can induce action (and we do know that advertising and propaganda "work"), which images might work best to raise awareness and induce remedial actions? As awareness of the impact of human agency on the world's climate increases, the sea is increasingly portrayed in complex and sometimes conflicting political memes. Most, however, are based on several deep fears related to the sea:

- *Fear of the Agency of the Sea.* In its "classical" mode as a force of nature the sea is portrayed as acting against humanity – water as active agent (tsunami), threatening images of beasts rising up, or destruction by icebergs;
- *Fear of the Depths.* In images of water as depth we can read a fear of drowning, of flooding, sea level rise, and monsters lurking beneath the surface; and
- *Fear of Humanity and the Anthropocene* – a more nuanced representation of humankind as agency itself against the planet.

Fear of the agency of the sea

In 2009, professor of visual culture Nicholas Mirzoeff posed an important challenge to the contemporary world: "What is the place of the sea in the human sciences? How can we interpret it as a material force and presence; as a place where power is marked and contested; and as a mythical or spiritual form of life that threatens humans and yet is also their vital support?"[11]

As recorded in creation myths around the globe, water played a central role in the creation of all life. In the beginning there was chaos, and out of the chaos, the gods created the known and orderly world, usually beginning with the seas. Even desert societies recognized the centrality of water in their stories. The biblical creation myth in Genesis adopted by Judaic, Christian and Muslim religions, states that in the beginning a "darkness was upon the face of the deep." In the Babylonian creation mythology, too, the goddess Tiamat represents the sea, "the deep," and chaos. Similar myths of water being present in the beginning are reflected in many other cultures: Iroquois, Egyptian, and Buddhist. According to Phelan, "The sea or the great waters, that is, are the symbol for the primordial flux, the substance which became created nature only having form imposed upon or wedded to it. ... The sea, in fact, is that state of barbaric vagueness and disorder out of which civilization has emerged and into which, unless saved by the efforts of gods and men, it is always liable to relapse."[12]

There is a term for this struggle with chaos, which has often been cast as a battle between nature and civilization. *Chaoskampf,* "the struggle against chaos," is "ubiquitous in myth and legend, depicting a battle of a culture hero deity with a *chaos monster*, often in the shape of a serpent or dragon."[13]

Underlying the fear of the agency of the sea, lies this deep dread – the primordial fear of chaos, of the void. The chaos is an uncontrolled and uncontrollable experience of endless space, immense scale, and vastness – as opposed to the intimacy and knowability of community, of place.

We know this as the sublime. Or even as hyperobject.

At the heart of our fear, of course, is the specter of death.

The sea not only represented a fear of chaos; deep respect for the sea co-existed with the fear of its agency and a deep respect for its ability to provide sustenance. In many cultures, the seas were a primary source of food and the sea was seen as the source of life. For example, the Inuit worshiped the sea goddess, Sedna, the source of the sea's bounty. The story of Sedna is rooted in violence – either against Sedna or on her behalf. In the most common versions, Sedna is thrown into the sea by her father. She clings to the side of a boat, but her father refuses to allow her to board, chopping off her grasping fingers. Her fingers became the life-giving creatures of the sea, and Sedna is consigned to the depths of the sea where she is viewed variously as a benign or extremely threatening sea goddess.

Agency suggests power and many artists have expressed this power through representation of the context of the sea. The ancients believed that the sea was alive, a great power. Further, it was filled with monsters seeking to destroy shorelines, boats, and seamen. Throughout human history, there were cautionary tales, and these myths and legends still haunt the human cultural psyche. But the power and the fear of the sea remained a sub-text – even in modern literature. Herman Melville captured the essence of the terrifying power of the sea in a passage from *Moby-Dick*: "No mercy, no power but its own controls it. Panting and snorting like a mad battle steed that has lost its rider, the masterless ocean overruns the globe."[14] He portrayed the ocean as a chaotic force that lurked as a sinister backdrop to the tale of the great white whale. Similarly, in the early twentieth century, the sinking of the Titanic caused by the collision with an iceberg served to remind humanity of the hidden power of the sea to overcome modernity. Initially, the sinking was blamed on faulty construction and human incompetence. More recent research has suggested that other natural forces may have played a role: atmospheric phenomena operated to move icebergs from their usual environment and create mirages that confused the ship's officers.[15]

How do we imagine the agency of water? Do we imagine it as a force of nature? Do we imbue it with consciousness and intentionality? In a broad discussion theorizing water's agency, cultural anthropologist Veronica Strang identified several elements of the agency of water, emphasizing the idea of flowing as dynamic and interactive, and as an expression of human/non-human connectivity based on the idea of "flowing between."[16] Jake Phelan also struggled with this question, as he attempted to enliven a sea that had become mere surface: "The sea is condemned as a blank environment, an empty space or void. ... And so I wish to emplace the sea, to create a space for it within the anthropolitical domain. Not by 'grounding' it, but by conceptualizing its very fluidity."[17] Jamie Kruse and Elizabeth Ellsworth identified several similar water-based agents: water as ice-shaper, water flooding, water flow, and the idea of water as embodying perpetual change.[18]

The idea of flow may be particularly useful in understanding the nature of the sea in contemporary cultural thinking. Instead of seeing the sea merely as surface or as idealized representation in a painting, flow implies being a part of something larger, having a relationship with, described by Australian cultural geographer Sue Jackson as similar to Heidegger's "living with," and Strang's idea of "flow between."[19] Flow, as an aspect of agency, also introduces the possibility of recognizing that the human being can be part of something beyond the self – the changing nature of the Anthropocene. "It may be said that the sea attracts by offering a force greater than the self. ... [T]he sea's movement and fluidity, as an uncontrollable force ... [that] has its own corporeal intentionality." An aspect of the "unknown, far off or deep down."[20] So, too, we are now figuratively "swept up" in nature.

Agency presupposes movement – and for our purposes, forceful movement. According to art historian Christine Guth in *Hokusai's Great Wave,* "Our visual perception of the sea is oriented toward its movement."[21] Guth explored the global and cross-cultural appropriation of Katsushika Hokusai's image of the "great wave" (the formal title of the work is "Under the Wave off Kanagawa," from Thirty-Six Views of Mount Fuji, 1830-1833), as it transitioned from an innocuous nineteenth century Japanese woodcut honoring a view of Mount Fuji, to its appropriation by commercial art as a global icon of tremendous energy, change, and motion. She also noted its role in political art as a representation of the threat of nature to humanity: "The notion of divine terror or the sublime ... further inflected Hokusai's great wave with notions of environmental danger."[22]

Guth also argued that Hokusai's great wave was used as a contemporary metaphor for the forces of globalization: "Because waves are in a perpetual state of flux, related to but not part of the shores on which they break, they are resonant signifiers of both the deterritorializing effects of globalization and the hybridity resulting from the mixing of their waters."[23] Guth further defined deterritorializing as the aesthetic process of freeing up this image of the great wave from its original context, and creating a free-floating meme that worked in many cultural and political contexts: "The visual language and subject of the great wave invite creative manipulation of its deterritorializing qualities to refer to or reflect on local sociocultural structures."[24]

Other researchers have cautioned against attributing intentionality to the sea. Nevertheless, according to Phelan, artists, particularly writers, have regularly assigned intentionality to the sea. Historically, writers of fiction were comfortable assigning malign agency to the sea: Homer, in *The Odyssey*, for example, describes a "cruel sea"; Joseph Conrad, noted its "unfathomable cruelty."[25] Throughout the novel *Moby-Dick*, Melville portrayed the sea as a dark menace, employing such terms as "cold malicious waves," "uncivilized," "omnipotent," and "everlasting terra incognita" to articulate the fearsome sea. The white whale, described as the biblical sea monster Leviathan, is not merely a beast of the sea, but an "intelligent malignity," the personification of evil. Melville brought the dark purposefulness and intentionality of the sea to life: "And heaved and heaved, still unrestingly heaved the black sea, as if its vast tides were a conscience; and the great mundane soul were in anguish and remorse for the long sin and suffering it had bred."[26]

Images of fear of the agency of the sea (and hence of nature) are portrayed in Figures 12-14. Katsushika Hokusai's iconic image of the great tsunami wave has been employed across many cultures to suggest the power of the sea. Russian artist Ivan Bilibin adapted the image in his illustrations for a Russian fairy tale, and Ricardo Levins Morales incorporated a tsunami to convey the power of a hurricane that struck Puerto Rico. In Homer's early Greek legend, the *Odyssey*, Odysseus had to sail between the perils of Scylla, a monster from the deep, and the churning seas of the whirlpool, Charybdis (not pictured). The whirlpool personified the agency of the sea and the powerlessness of humanity. Both elements of nature were intent upon drowning Odysseus.

The seas have inspired other fears – the ancient fear of the flood in the biblical tale of Noah and the ark is probably the most iconic narrative conveying the threat of being overwhelmed by water, and even a personal fear of drowning are embedded in the visual narratives. According to Mirzoeff, "the flood is perhaps the oldest metaphor that humans have preserved in first, Mesopotamian, and then Jewish legend. Indeed, the ancient god Yahweh was a storm god, capable like the mysterious flood of manifesting anywhere and anytime."[27] Ovid's *Metamorphosis* includes an early Roman creation story – a tale of global inundation and the subsequent re-population of the planet by two surviving humans. The ongoing searches (even to this day) for the lost city of Atlantis that was swallowed by the sea, the tales of ships disappearing in the Bermuda Triangle, and the unimagined horrors of the flooding of New Orleans during hurricane Katrina and the destruction visited on the Bahamas by hurricane Dorian, attest to this unease and the ongoing fear of the power of water. It is interesting to note that hurricanes are named, thus unconsciously perhaps assigning them being and hence the possibility of conscious intent. The contemporary environmental instantiation of this fear is, of course, rising sea levels.

Fig. 12 (above): Katsushika Hokusai, *Under the Wave off Kanagawa*
Fig. 13 (below): Ivan Bilibin, *Illustration for* Pushkin's Tsar

Fig. 14: Ricardo Levins Morales, *San Ciriaco 1899* Puerto Rico hurricane

Metaphors for flooding and inundation are common in human language and art. A "flood" is often referenced to describe anything that can overwhelm human beings: a flood of tears, a flood of job offers, emotions, or memories. In contemporary fear-mongering politics, there are multiple references to a "flood" of refugees overwhelming Europe and the United States – all designed to resurrect and trigger the ancient connection with the fear of water.

With the voice of an ancient prophet, Melville admonished humanity to beware of the power of the sea. The fear lingers in the cultural stories: "Yea, foolish mortals, Noah's flood is not yet subsided; two thirds of the fair world it yet covers."[28] Today, warnings of rising sea levels are increasingly common, and hurricane predictions now come with estimates of the height of storm surges, again producing anxiety and concern. The threats of inundation are regularly conveyed by maps showing how coastlines will recede, which cities will be overwhelmed, and dire predictions that Venice and Miami will succumb to the sea. Dystopian narratives are often set in a watery environment like the film *Water World* and Kim Stanley Robinson's novel, *New York 2140*. Despite these dire warnings and high tides that regularly flood downtown Miami, Florida streets, the city continues building along the seashore.

Fear of the Depths

It was not only the surface activity of the sea and the dangers emerging from its depths that instilled fear. Throughout human history, thalassophobia was also fed by fear of unknown creatures living below the sea – some real, some imagined, most unseen: the biblical Leviathan, the octopus (the Norwegian kraken), the Japanese Umibōzu who wrecked ships on the sea [Figures 15 and 16]. Herman Melville captured this sense of dread in a passage describing nature as a great force. The creatures of the deep hidden beneath the surface symbolized a lurking sense of evil.

> Consider the subtleness of the sea; how its most dreaded creatures glide under water, unapparent for the most part, and treacherously hidden beneath the loveliest tints of azure. Consider also the devilish brilliance and beauty of many of its most remorseless tribes, as the dainty embellished shape of many species of sharks. Consider, once more, the universal cannibalism of the sea.[29]

Fig. 15: Pierre Denys de Montfort, 1801

Any number, shape and size of sea monsters (of which there were reports and well documented sightings up until the nineteenth century) were the everyday embodiment of a great terror rising up out of the sea. The fear of sea monsters even persists into the modern era. Since the 1950s the Japanese *Godzilla* movie franchise has terrorized several generations worldwide, by capitalizing on the global fear of nuclear contamination that would return from the sea in horrific form to complete the destruction of civilization. The shark, too, has come to represent this terror of the depths – a fear based in reality. In *Fish Story*, photographer Alan Sekula pointed to the movie *Jaws* as shifting contemporary ocean iconography from focusing on the surface to generating fear of the depths.³⁰

Fig. 16 Utagawa Kuniyoshi, *Umibōzu* 1843-1845

The threat of rising seas is subtly but effectively conveyed in Figures 17-22. Andrei Popov's *Climate Change* articulates visually the anxiety about the futility of human efforts to "nail down" the sea level. Riber Hansson's image of an overcrowded world as ark in a flooded planet brings to mind the biblical terror of the Great Flood. Leo Lin's image is a simple, straightforward image tagged with the phrase, "Global Warming," and Scott Laserow's moving image of hurricane Katrina suggests both drowning and the loss of New Orleans. Kiakili's powerful image of a black man drowning, threatened by guns and either indifferent bystanders or supporters snapping photos, is an excellent example of what Guth called "de-territorializing" iconic images – that is, using them in a different context but continuing the power of the icon. The stark reality of death is brought home in this image, and Sam Kerson and Katah's, *Drowning*, which recorded in real time the tragic deaths of refugees trying to reach the safety of southern Europe's Mediterranean shore.

Fear of humanity
imagining the Anthropocene

What are we to make of the human impact on the planet as a whole? Here, too, artists are using the sea as a powerful icon portraying the Anthropocene and the destructive impact of humanity on the planet. In previous sections,

Fig. 17: Andre Popov, *Climate Change*

Fig. 18: Riber Hansson, *The Ship Earth*

Fig. 19: Leo Lin, *Global Warming*

Fig. 20: Scott Laserow, *Hurricane Katrina 2005*

Fig. 21: Kiakili

Fig. 22: Sam Kerson and Katah, *Drowning*, 2016

the agency of the sea (and the creatures under it) was portrayed as an action by "nature." In this section, the sea will take on a new role: as icon for human actions against the environment, by integrating fear of the agency of the sea with the consequences of human activity. Mirzoeff noted that "the Anthropocene is a human-created machine that is now unconsciously bent on its own destruction, a purposiveness without purpose."[31] By its very definition the term Anthropocene requires that we address human agency in global warming. One writer has suggested we call the phenomenon ecocide.[32] Human response ranges from denial of the reality of the Anthropocene, to actively visualizing the force of humankind's impact. How, then, are we to understand the actual agency of human activity and how to communicate when that agency is met at every turn by non-human agency? According to Mirzoeff, "Since the seventeenth century … the modernizing call for the 'conquest of nature,' a visualization of the planet as an enemy to be subdued," has driven contemporary images of the sea, and thereby influencing how we might alternatively imagine the Anthropocene.[33] If the planet is viewed merely as property, as object to be appropriated and colonized, it becomes difficult to imagine it as having equal weight and force, or "being with" nature, even when that relationship reveals our abuse. The challenge is how to visualize or counter the human denial and the indifference, and change the narrative.

One of the most powerful visualizations of the consequences of humanity's environmental hubris can be found in "The Sorcerer's Apprentice" sequence in the 1940 Disney film *Fantasia*. The apprentice takes up the magic wand of his master and calls upon its magical power to automate the tiresome task of carrying water by ensorcelling a broom to do the work for him. He falls asleep, and awakens to being inundated by a great flood of water. He frantically chops the broom in pieces, but each piece re-forms as another broom, relentlessly carrying even more water – nature out of control. In the story, the apprentice is saved only by his master's return and the reversal of the waters. But perhaps a more meaningful lesson can be drawn from this powerful tale in the new era of the Anthropocene. All human sorcery is ultimately powerless in the face of overwhelming reality of natural forces. While this tale leaves the elder sorcerer still in a position of power – it is merely another example of the false magic of hubris.

According to Indian novelist Amitav Ghosh, we are, despite our so-called modern existence, actually very vulnerable to the natural world. It is therefore imperative that we recognize (re-know) the human interrelationships with nature to ensure our continuing existence. Ghosh noted that "an awareness of the precariousness of human existence is to be found in every culture," in myths and legends around the planet.[34] Modern humans have ignored these time-bound insights to our great peril. Mirzoeff also alluded to this precarity: "What is ultimately far more disturbing to modern thought is the potential realization that no one (or nothing) in fact has authority."[35] In the modern era, the awareness of our precarity in the face of nature has been suppressed, as objective scientific explanations replaced older beliefs and fears. "The victory of gradualist views in science was similarly won by characterizing catastrophism as un-modern. ... I suspect that human beings were generally catastrophists at heart until their instinctive awareness of the earth's unpredictability was gradually supplanted by a belief in uniformitarianism."[36] [Ghosh, 22, 25] There was no room in the modern consciousness for a wild and uncontrolled nature, let alone the catastrophe of chaos. But respect may be based on a positive respect for fear. Ghosh concluded that we really cannot escape this awareness, the power of non-human agency. The power of the sea can serve to make us aware of our vulnerability by representing catastrophe. The Anthropocene – human agency – is itself, alive, and the agent of the great forces moving against the planet:

> [T]the freakish weather events of today, despite their radically nonhuman nature, are nonetheless animated by cumulative human actions. In that sense, the events set in motion by global warming have a more intimate connection with humans than did the climatic phenomena of the past – this is because we have all contributed in some measure, great or small, to their making. They are the mysterious work of our own hands returning to haunt us in unthinkable shapes and forms."[37]

Philosopher Arnold Berleant invited us to return to the metaphor of the sea as a way of imagining a renewed relationship with the natural world. Water has much to teach us as we struggle to find a new way of thinking about the Anthropocene: "A different metaphor would serve us better, one that points up the force, pressure, interdependence, and continuity that are central to environment, both understood ecologically and experienced aesthetically. That metaphor is provided by water."[38]

> [T]he water environment forces us to see ourselves as an inseparable part of those processes. We are immersed in the world, which is at the same time a world transmuted by human agency. ... [W]ater environments lead us to respect natural processes. They give us a sober sense of human proportions and limitations. Because water environments are largely not human-made and so are not in the image of human culture, we are forced to recognize the limits of our power. Recognizing this with our bodies as well as with our understanding is a profound

environmental lesson. To live, then, as a harmonious part of the natural process is to be most truly human.³⁹

Figures 23-24 which incorporate images of a tsunami, address the reality and force of human agency AND serve to break down the inner/outer, subject/object dichotomy. There is no nature "out there" – human agency is seamlessly merged into these tsunami images, and is integral to the images. The tsunami is not acting with its own agency, rather embedded in these images is human agency. We have become tsunami.

Conclusion

Fear of a tsunami or rising sea levels is a high-fear event, so we would have to ask what would induce us to pay attention to these high-fear inducing images. Advertising effectively employs low-fear tactics – which work, but only in a general, vague manner. High-fear tactics, however, are generally not used in advertising. A little social anxiety might be a good thing, but deep fears are not likely to induce people to buy something or take action. It seems to me that selling a product calls for a "good news, problem solved" experience – the minor fear can be easily addressed and the solution is quick and easy. But environmental fears are of an entirely different magnitude. They call forth not just anxiety, but fear for survival, reaching us at the most visceral level. For this reason, the high-fear tactic seems appropriate. We pay attention when the tornado sirens

Fig. 23: Bill Roberts

Fig. 24: Sebastian Thibault

broadcast their warning to take cover, don't we?

As we have seen, images of fear of the sea in political and environmental art fall into several categories: a giant tsunami overtaking humanity; sea monsters rising up and attacking humanity; human vulnerability portrayed by drowning in rising seas; or even the sinister actions below the sea by icebergs. Emerging memes are more sophisticated. They portray the forces of water as agents of the Anthropocene but they bring to the story the impact of human agency threatening not only the environment, but humanity itself. They use, either actively or passively, the fear of the power and agency of the sea as it intersects with human survival, putting humans into the picture as co-partners with the natural world. They tell us that we are the cause of the pending catastrophe and we should be afraid.

But do these images work to alert you to the multiple threats of the Anthropocene? What do you think? Only time will tell.

NOTHING WILL EVER BE THE SAME AGAIN
QUESTIONING POLITICS AS USUAL

"Of course we need hope. But the one thing
we need more than hope is action. Because
once we start to act, hope is everywhere."
– Greta Thunberg

The climate crisis is a liminal space. There are many ways to describe the liminal state. It can be the place where the ocean meets the land, an ever changing environment of movements both great and small – where the sea erodes the shoreline and the shoreline resists the sea; where plants and animals are constantly adjusting from environments of liquidity to solidity.

The liminal can also be a threshold (the source of the word) of opportunity – a door opens and we must step through and decide which way to go. In the field of anthropology the concept of liminality is used to describe a rite of passage – the transition from one state of being to another. We might even call it a kind of twilight zone. Liminality can also describe a time of profound social change; a period of uncertainty, anxiety, and fear; of constant changes;

and especially of precarity – being dependent on the will or favor of others resulting in instability.

Is this sounding familiar to you? The climate crisis is just such a liminal moment, betwixt and between, hovering on the edge of an unknown abyss, waiting on the verge to decide which way to move. We have no idea what shape the climate crisis will take tomorrow, and we generally have no control over how it will proceed, David Wallace-Wells argued: "[U]ncertainty is among the most momentous metanarratives that climate change will bring to our culture over the next decades – an eerie lack of clarity about what the world we live in will even look like, just a decade or two down the road. ... [T]hat haunting uncertainty – emerges not from scientific ignorance but, overwhelmingly, from the open question of how we respond. ... What will we do?"[1]

As the world as we know it begins to unravel, what will we do? Will we cringe paralyzed in the corners of our lives, or will we seek new ideas and creative alternatives to manage our lives differently? There are many possible paths we could choose: collapse in terror and isolation, take up arms against one another, long for a superhero to save us, or join with others to begin to address the many problems we will be facing and the actions needed or goals to set.

Politics in the Anthropocene will be very different. When – not if – the capitalist economic system with its embrace of fossil fuels, continual growth, and endless consumption is severely curtailed, we will need to invent new political and economic structures to implement our collective goals. Very briefly, capitalism is the enclosure of previously commonly-held resources via private property rights to ensure the privatization of everything to profit individuals and corporations. Capitalism is based on several principles: the exploitation of labor (you do the work, but do not benefit from the profit); the exploitation of diminishing natural resources to continue generating wealth; the accumulation of wealth ("capital"); continued economic growth (and thus continued exploitation of labor and the environment); the socialization of pollution costs requiring the taxpayers pay to clean up the messes; the profits going to individuals and stockholders rather than the workers who made the product; and the encouragement of excessive consumption through stimulation of human desires to create continued growth (that's you and me who fulfill our deepest longings by buying all this stuff). Capitalism is not designed to meet the basic needs of all – it is designed to keep the system of perpetual accumulation running in the hands of a small percentage of individuals and creating ever increasing profits for them at the expense of the majority of humankind and the planet.

In this chapter and the next I will explore various options for how we might proceed to bring some order to our lives in an unstable, changing environment where our politics, our economics, and our relationship with the natural world will have to be rethought in collaboration with the realities of a changing natural environment.

We are all connected to the life and well-being of the planet. We now have a term for human agency and our destructive interactions with nature – the Anthropocene. We have become destroyers, perpetrators of violence against nature and ourselves. We have yet to come to grips with the reality that we, too, as a species, are endangered, as Dipesh Chakrabarty explained:

> [U]ltimately, what the warming of the planet threatens is not the geological planet itself but the very conditions, both biological and geological, on which the survival of human life as developed in the Holocene period [prior to the Anthropocene] depends. The word that scholars such as Edward Wilson or Paul Crutzen [who first coined the term Anthropocene] use to designate life in the human form – and in other living forms – is species. … Species thinking … is connected to the enterprise of deep history. Further, Wilson and Crutzen actually find such thinking essential to visualizing human well-being.[2]

We know the old ways of doing business won't work. The only option is to join with the planet for mutual benefit. Nature must have a seat at this table. It will be the ultimate political act – all life together taking direct action to save all life. The capitalist-driven idea of progress is part of the problem and cannot be the solution; technology alone won't save us; the rapture isn't happening; and government, politicians, and corporate leaders have been notoriously reluctant to act. Despite all the investments by the wealthy in space travel, we cannot escape to another planet. We're stuck here, so we have to roll up our sleeves and get busy.

There are two ways to think about a human response to how human beings have created the conditions of the Anthropocene: as a fatal flaw in the biology of human beings (our species nature made us do it), or as a function of our human intellectual efforts – the consciously designed technological, political, and economic forces underlying humanity's relationship to the world. If it truly is a result of our biology, there is little we can do about our species nature – we can't really "act" to change our very being. If that is the case, all we can do is sit back, put our feet up, pop open a can of beer, and watch the world disintegrate around us. Blaming the Anthropocene on our species nature is only an excuse to fail to act, I believe. According to professor of cultural studies Heather Davis and Etienne Turpin, "[F]iguring the Anthropocene as a 'species question' hides the most significant problem of our present situation: the asymmetrical power relations that have resulted in the massive transformation of the Earth through industrialized agriculture, resource extraction, energy production, and petrochemicals."³ Instead of this excuse, which misses the crucial role of how humans have perceived themselves as the dominant force in nature, we can focus on human intellect and develop responsible solutions to overcome, amend, and repair humanity's destructive political and economic activities. There is a small window of time in which we can undertake conscious actions to make the necessary changes and move beyond the elephant in the room – capitalism – to create a new world.

I have divided the discussion of politics in the Anthropocene into two chapters. This chapter will critique several options that essentially leave current political and economic structures in place with an eye to reforming environmental practices. Now that we know our actions have changed the course of geologic history, we can, it is assumed under this scenario, begin making conscious choices to reverse the damage. We are facing a future of profound upheaval and uncertainty, and like most people, I would like things to remain as stable as possible but it is increasingly clear that we need to plan for tremendous changes, and to take initial remedial actions to reduce fossil fuel usage as the threats to human survival grow. In the past 100 or so years we have been burning millions of years of sequestered carbon back into the atmosphere, and this cannot continue if we expect to keep the temperature of the planet within habitable range for human beings.

We are at a point now where decision-makers know we have a catastrophe on the horizon. Indeed, there is already enough buy-in from forward-thinking governments and businesses that mitigation plans, usually of a technological nature, are underway on many fronts. Scientists and engineers in particular are busy exploring and re-examining new energy sources like cold fusion technologies, re-designing data storage that dramatically reduces energy consumption by the computing industries, and proposing geo-engineering projects. Most of these efforts implicitly assume the maintenance of current lifestyles and economic models. Whether these actions are sufficient remains to be seen. Soon the debate will sharpen and harder choices will have to be made.

According to professor of global change science Simon Lewis and geographer Mark Maslin, in *The Human Planet: How We Created the Anthropocene*: "There are just three possible futures: continued development of the consumer capitalist mode of living towards greater complexity; a collapse; or a new mode of living." The first future they described as "a path of business-as-usual with reforms." The second – collapse – concluded that the environment will overcome any efforts, creating catastrophic fallout worldwide, a collapse "that would still likely take the form of private property and a free-labour-based capitalist mode of living. … this option may also seem probable." Their third option is the creation of a new system replacing industrial capitalism that will demand greater energy consumption, more information processing, and more collective efforts. They are not confident that our current capitalist/

consumerist system will prevail, leaving only the option of collapse or a shift to a new model.[4] Will we continue down our current path of rampant consumption or downsize to fit within nature's parameters?

The next chapter explores more radical survival-oriented local and collective actions – a restructuring of personal and political activities that we can start implementing even today. All of the existing systems of economics, social organization, and politics are likely to fail in one degree or another. In case I'm right, we need to be prepared to create the kind of society that can function in a revised relationship with the other beings on the planet by considering radical alternatives and the likelihood of austerity as a permanent reality. By staying the current course with only minimal disruptions we run the risk of underestimating the looming forces of the natural process we have set in motion.

How Will We Approach the Future?

There are many pathways we can take to achieve the goal of co-habitation with the natural world and the other beings with whom we share it. How much our lives will be upended, how quickly we react to change our ways, and how seriously the environment will be damaged, are all variables that will determine what decisions we make. But before we do anything, our future must begin in acceptance, grief, and obligation. We have to come face to face with the results of what our actions have set in play in the world, the human activities that have brought us to the edge of the precipice of self-destruction. Grieving the damage we have done to the environment is a necessary part of moving forward, according to Timothy Morton: "The only firm ethical option in the current catastrophe … is admitting to the ecologically catastrophic in all its meaningless contingency, accepting responsibility groundlessly, whether or not 'we ourselves' can be proved to be responsible."[5] We can begin by coming to terms with the grief that we feel, and accepting the responsibility for our collective damage, so that we can willingly take up the tasks ahead of us. Grief is not despair – it is a process that allows us to move ahead a bit sadder and hopefully much wiser. Beyond grief is action. No more denial, no more thinking we will be the exception to the harsh realities we must face, no more hoping that a god or a leader will save us.

> Now is the time for grief to persist, to ring throughout the world. Modern culture has not yet known what to do with grief. Environmentalisms have both stoked and assuaged the crushing feelings that come from a sense of total catastrophe, whether from nuclear bombs and radiation, or events such as climate change and mass extinction. … If we get rid of the grief too fast, we eject the very nature we are trying to save.[6]

Taking responsibility will not be easy. In an interesting article exploring the psychological anguish and philosophical trauma experienced by people who have accidently killed other people, writer Alice Gregory discussed Bernard Williams' concept of "moral luck," a theory exploring the relationship between culpability and responsibility. The response/question going forward should be, yes, I did it, but in what way am I responsible for it? According to Gregory, "Williams defined the torment of the person who causes an accident as 'agent-regret,' a type of first-person remorse that is distinct from that of a mere bystander. ... He suggests that we need a more nuanced understanding of human agency, which acknowledges that one's history as an agent is a web in which anything that is the product of the will is surrounded and held up and partly formed by things that are not.'"[7]

It seems to me that there is a way to deal with the awareness of human destruction of the environment without the individual succumbing to self-recrimination resulting in paralysis. We all – through both individual and collective actions – have trashed the planet, some more than others (yes, I'm pointing my finger at you, United States and Europe). How, now that the deed is done, do we come to terms with recognizing the damage we have done and our individual and collective responsibilities to make it right? The political battle is ultimately between two world views – will we choose the anthropocentric (human centered) perspective or the ecocentric (earth centered)? Our responses will flow from the path we choose.

The climate crisis will occupy our attention for decades to come. Taking responsibility allows us to move beyond the paralysis of hopelessness to political action. Instead of despair, hope can result from fulfilling our obligations. In responsibility we will also find empowerment. The economic and political system we currently live in works every day in every way to encourage dependency and undermine our self-confidence and our self-determination to respond to the world. Uncovering our own possibilities will prove to empower us even further. As philosopher Nathan Jun has observed, "That system [capitalism] reproduces itself in large part through its politics – a politics of despair that thinks the possible via the actual. Libertarian socialism, in contrast, aspires to demonstrate that the actual is, in fact, the denial and suppression of the possible. It is a politics of hope that seeks to 'possibilize' the actual, but it is also a politics of truth that uncovers an immanent lived (and) living reality just beneath the veil."[8] [Personal correspondence, June, 2019]

Change Will Be the New Air We Breathe

As we watch previously predictable weather changes become collapses of whole climate systems, or as the waters of the ocean slowly swallow up city blocks, or as crops fail from years of drought or too much rain, and forests burn out of control, one of our big challenges will be to bring to the whole planet our belief in the power of people to embrace change. But let's face it, we don't really like change. We like things to remain stable, we like the predictability of knowing (at least sub-consciously) that tomorrow will be pretty much like today. But change is coming to our world, whether we like it or not, and whether we are ready or not. Hope lies in moving more consciously to position ourselves and our communities for action in light of even more changes coming.

Our future success requires that we be open to changing quickly, despite the uncertainty about how the Anthropocene will progress. Changes to the environment seem to be occurring at a faster pace, such that scientists are having to continually re-adjust models and projections. We need to be flexible. To complicate things further, the changes are not likely to be linear or predictable. As sustainability scientist Jem Bendell noted, "Non-linear changes are of central importance to understanding climate change, as they suggest both that impacts will be far more rapid and severe than predictions based on linear projections and that the changes no longer correlate with the rate of anthropogenic carbon emissions." In other words – "runaway climate change."[9]

We will have to learn to be more nimble on our feet, and more open to adapting our life plans. Our very survival will depend upon it, according to Kathryn Yusoff and Jennifer Gabrys: "[O]ur ability to imagine other possibilities, to embrace decidedly different futures with creativity and resolve, to learn to let go of the sense of permanence we may have felt about certain landscapes that have seemed to be always so, and to embrace change, is paramount to building resilience and adaptive capacity."[10]

What About Capitalism
Can't the Free Market Fix Things?

Let's begin by examining briefly the current political and economic structure of our society. Some writers have argued that the Anthropocene should, in fact, be called the Capitalocene – after all, capitalism is the dominant global economic and political system that initiated and continues to drive the destruction of the planet.

The fuel running this economic engine is fossil fuel, the runaway polluter of the environment.

There is wide-spread agreement that capitalism and the planet are on a collision course that is already well under way. This chapter will not spend time making the case against capitalism. If you've been paying attention, you already know that it is at the core of the environmental catastrophe. The case against capitalism has been well-argued by the likes of Naomi Klein's *The Shock Doctrine* and *This Changes Everything*, Joel Kovel's *The Enemy of Nature*, Bill McKibben's works, among many others too numerous to mention. Even the most cursory of Google searches will bring up tens of thousands of references. Our challenge will be to quit wasting time and start focusing on ways that we can undermine and disarm the economic forces that are of our own making, and were originally embraced and celebrated with our warm approval. We are complicit, and the response must reflect this recognition. As professor Michael Truscello argued in a radical critique of Naomi Klein's book *This Changes Everything*, "The idea that humanity could fight climate change without dismantling the ecocidal system that created climate change is the kind of logic eco-opportunists engage when trying to lower the bar for environmental activism."[11]

What about This Idea of Sustainability?

Calls for "sustainability" have dominated recent economic discourse about how to respond to the Anthropocene. What "sustainable" actually means in real, material terms, however, is seldom explored. To date, neo-liberal circles have concentrated on assuring us that capitalism can maintain the current level of production and consumption, and with some tweaking mitigate the damages to the environment. However, Timothy Morton reminded us that talk of sustainability might be a wolf in sheep's clothing: "An awful lot of ecological speech is actually oil economy speech. In fact, almost *all* ecological speech isn't ecological speech at all. Ecological speech is deeply distorted by the oil economy we live in. All that language about efficiency and sustainability is about competing for scarce highly toxic resources."[12]

The idea of sustainability is embedded in recent proposals in the United States for a Green New Deal (GND) – a plan that is clearly designed to keep capitalism alive. Capitalists will continue to work hard to normalize and (of course) monetize the climate crisis. As I write, corporations are ramping up the extraction of any number of natural resources: trees, fossil fuels, minerals – but in a supposed safe and earth-friendly manner. Writer Wayne Price summarized the many problems with the Green New Deal:

> [T]he Green New Deal strategy is problematic because it means [1] an effort to modify existing capitalism, not to fight it with the aim of overthrowing it. [2] As often stated, it requires working through the Democratic Party. [3] It proposes to use the current national state as the instrument of change. [4] Finally, while advocates speak of popular mobilization and democratization, their overall approach is top-down centralization.[13]

A core proposal of the Green New Deal calls for decarbonization of the atmosphere to mitigate the underlying cause of global warming. A study from

2017 noted that the United States and the European Union together, have contributed over 50 percent of CO2 emissions since 1850. Their debt to the rest of the world is significant. The World Economic Forum explained how decarbonization is supposed to be implemented globally.

> According to the scientific consensus, climate stabilization requires full decarbonization of our energy systems and zero net greenhouse-gas emissions by around 2070. The G-7 has recognized that decarbonization – the only safe haven from disastrous climate change – is the ultimate goal this century. And many heads of state from the G-20 and other countries have publicly declared their intention to pursue this path.
>
> Yet the countries at COP21 [signers of the Paris Agreement] are not yet negotiating decarbonization. They are negotiating much more modest steps, to 2025 or 2030, called Intended Nationally Determined Contributions (INDCs). The United States' INDC, for example, commits the US to reduce CO2 emissions by 26-28%, relative to a 2005 baseline, by 2025.[14]

Note that full decarbonization is the ultimate and necessary objective and all the politics between now and the end of the century and beyond have to be directed at achieving this goal. So far, each country determines its own "intentions." There is no United Nations monitoring for compliance, no coordinated commitments. We also need to remind ourselves that none of these agreements are as yet binding. Unfortunately United States President Donald Trump has withdrawn from the Paris Agreement, backing away from even the very modest commitments proposed by the United States above.

Geographers Simon Lewis and Mark Maslin were very skeptical about decarbonization as a viable means of removing carbon from the atmosphere and thus limiting the increase of carbon dioxide which is raising global temperatures. These plans do not challenge the core causes of carbon in the atmosphere but rather argue for greater regulation, more sophisticated technology, and monetizing carbon pollution. In a detailed discussion of various technological proposals for Bioenergy Carbon Capture and Storage [BECCS] currently under consideration, they warned of the dangers of taking this direction to mitigate current economic carbon production. It would require vast amounts of land to build the thousands of plants necessary to extract the carbon dioxide from the air and bury and permanently store it – "one to two times the size of India," and "the current plans of the world's governments, if followed, will 'save the world' by destroying it another way. ...The underlying reason BECCS is so attractive is because it puts off taking action now. ... The delay-climate-action-and-make-nature-pay-later story is not a wise one to tell ourselves. In essence it is still the old religious idea of human dominating nature rendered in mathematical equations."[15]

Ecosocialist writer Richard Smith has also called the whole concept of decarbonization into question because it fails to consider the necessarily radical implications of dramatically changing the energy economy. We live in an oil economy and serious alternatives have yet to emerge.

> What's not said is that decarbonization has to translate into shutdowns and retrenchments of actual companies. How does one decarbonize Exxon-Mobil or Chevron or Peabody Coal? To decarbonize them is to bankrupt them. Further, the same is true for many downstream industrial consumers.
>
> Perhaps the biggest weakness of the GND Plan is that it's not based on a fundamental understanding that an infinitely growing economy is no longer possible on a finite planet…, no acknowledgement of the imperative need for economic degrowth of many industries or of the need to abolish entire unsustainable industries from toxic pesticides to throw-away disposables to arms manufacturers.[16]

If capitalism must be deactivated, how can we in good conscience, as Smith asks, promote electric vehicles to replace gasoline engines, by overlooking the fact that power plants also consume fossil fuels? Cutting back – a good option – will not contribute to profits either, so don't look for any support from shareholders for a message of reducing consumption.

Decarbonization plans do not address overall consumption and its fallout either. We already know that solid waste production is steadily increasing and recycling is an inadequate response, especially since the United States can no longer export its recyclable garbage to other countries. The most recent data from the Environmental Protection Agency (EPA) on recycling in the United States is from 2017. That year, a total of 267.8 million tons of municipal solid waste [MSW] was produced (and continues to grow), of which 67.2 million tons was recycled and 27 million tons was composted. Most of the recycled material was paper products, much of which was burned and hence returned more carbon dioxide to the environment (34.02 million tons). The EPA referred to this burned recycled material as "energy recovery, totally ignoring its impact on the atmosphere.[17]

Beyond Business as Usual

If we are truly committed to making the necessary changes to our economy that will allow us to live within the parameters of the natural world, capitalism as it is currently configured will no longer be viable. Along with a general call to "do no harm" as we seek solutions to preserving the environment, we might also consider adopting the recommendation of The Wingspread Statement to guide all future environmental planning: "When an activity raises threats of harm to human health or the environment, precautionary measures should be taken even if some cause and effect relationships are not fully established scientifically" (*Wingspread Statement*, 1998). The roots of this principle can be traced to the writings of environmentalists Aldo Leopold and Sir Austin Bradford Hill.[18]

There are, in the end, actually only a few core political and economic decisions to be made – all difficult. It will be necessary to mitigate the most egregious damages to the environment by reducing carbon in the atmosphere; reducing energy consumption overall and expanding renewable energy technologies; shifting from individual cars to public transportation; improving agricultural practices; and reducing our consumption of meat and dairy.[19] The central question for politics is, of course, how will these goals be achieved? That depends, of course, on who holds political power and whether or not they even believe in the climate crisis. Our challenge will be to figure out how best to move forward.

If we choose to leave most current systems in place, we can at least begin the process of decoupling our lives from destructive systems by examining the several options outlined in the following sections. Studying these options will help to clarify how we could more definitively frame and operationalize alternative solutions. The options range from tinkering with minimal efforts, to serious planning for adapting to the changing climate, to gradually accommodating in a downward spiral to an increasingly degraded environment. There are key questions of a political nature to be addressed as we study the various options for how to proceed.

Should we freeze economic activity to the current level of consumption? Or should we explore a significant downsizing of the economy? Where should decision-making take place? Some efforts will require international or nationwide agreement and coordination, others can be taken up by states, regions, local communities, small groups, or even individuals. The federal government encourages lower automobile fuel emission standards? California establishes its own standards and continues to require more fuel-efficient vehicles despite the relaxation of federal guidelines. Government won't invest in solar panels? Why not a neighbor cooperative sharing of expenses to buy your own panels for local collective use. Government won't regulate fuel emissions? Why not cooperate with neighbors to reduce fossil fuel consumption by setting up carpools or collective ownership of a car? Dietary changes? Deindustialize farming which is heavily fossil fuel dependent, reduce beef consumption, buy locally produced goods, and return to backyard and community gardens. Do we really need bananas from Central America to meet our nutritional requirements? A further benefit of giving up bananas would be the return of banana plantations to local populations, allowing them to produce other foodstuffs for their own neighbors and families. Best of all, we don't have to wait for Washington D.C. to make these important decisions – we have the option of generating local and individual decisions.

Option One
Essentially maintain the current status of the global economy

There is a powerful, but conservative, argument to be made that neo-liberalism (i.e., the political philosophy of consumerism and free market capitalism) will offer the best likelihood of addressing the problems that neo-liberalism itself created, according to David Wallace-Wells. [20] He outlines four possible models developed by Geoff Mann and Joel Wainwright in *Climate Leviathan: A Political Theory of Our Planetary Future* that nation states could adopt to maintain a capitalist system both politically and economically. The models differ in their emphasis on national versus global political control and national politics versus the interests of capital. Wallace-Wells concludes that China will likely end up in a global leadership role because it is the largest and richest economy in the world.

There are several real dangers, however, in trusting that current global economic and political systems will actually solve the environmental problems – especially if the profit motive continues to drive decision-making. Some of these dangers include how the current economic system will do everything in its power to shape our thinking, how other nations will react, or how we could be sold alternatives that will only create a different set of problems. According to Uri Gordon, some of these intemperate responses might include:

- The normalization of environmental and resource crises,
- The commodification of the atmosphere, as marketable debt [e.g., the buying and selling of carbon],
- The re-branding of nuclear energy as a "clean" alternative to fossil fuels, and
- The absorption of ecological consciousness into consumer culture via new organic food and clothing markets, "green" shopping malls, and the personal carbon offsetting industry.[21]

Option Two
Adaptation

This option calls for a gradual and careful adjustment to the impending disruptions caused by global warming and assumes an orderly transition through better management. It leaves basic political and economic systems in place, emphasizes local economics and adaptation to the environmental changes in that locale, and establishes in the process a baseline of survival needs. Planning would anticipate where system breakdowns make people most vulnerable, recommend sharing resources, make decisions about what to discard (does every single family need its own washing machine, dryer, car, snowblower), or even decide what we might not need at all. An advocate of the adaptation position is American journalist Chris Hedges, who argues that "We must invest our energy in building parallel, popular institutions to protect ourselves and to pit power against power. These parallel institutions, including unions, community development organizations, local currencies, alternative political parties and food cooperatives, will have to be constructed town by town. The elites in a time of distress will retreat to their gated compounds and leave us to fend for ourselves."[22]

In a widely shared white paper, "Deep Adaptation: A Map for Navigating Climate Tragedy," British sustainability expert Jem Bendell develops a series of recommendations he call the Deep Adaptation Agenda to guide the response to global warming. In the paper, he outlines a variety of programs already underway, including a goal built into the Paris Agreement in 2015 and the United Nations agency on disaster reduction, to support adaptation efforts worldwide that promote "'resilience', rather than sustainability." Bendell nevertheless recommend that we proceed with caution in adopting the Paris objectives. "First, the upbeat allegiance to 'development' and 'progress' in certain discourses about resilience may not be helpful as we enter a period when material 'progress' may not be possible." A second problem is that the programs focus on changes to the physical environment and overlook psychological resilience as an objective, which may actually be more useful helping people adapt.[23]

Bendell then takes up the sensitive topic of "relinquishment," or what I would call large-scale downsizing coupled with reduced expectations: "It involves people and communities letting go of certain assets, behaviours, and beliefs where retaining them could make matters worse."[24] This will likely be the most controversial and contentious recommendation for modern consumer cultures to adopt. In the United States, lowering expectations is already being met with fear and anxiety over real and perceived falling standards of living. As the recent election of Donald Trump has shown, a violent conservative backlash could result in response to mandated lower expectations.

Bendell's third recommendation for deep adaptation is to begin a process of "restoration," where we return to more earth-friendly behaviors: walking instead of driving, letting our lawns grow, eating locally and seasonally, gardening. Journalist Dahr Jamail and artist Barbara Cecil concluded that Bendell's proposals should not be read only as limiting our futures. They can open up new ways of accommodating ourselves to future challenges:

> Adaptation is new territory. Here is the realm of healing, reparation (spiritual and psychological, among other ways) and collaboration. It is strangely rich with a new brand of fulfillment and unprecedented intimacy with the Earth and one another. It invites us to get to the roots of what went astray that has led us into the sixth mass extinction. Given that with even our own extinction a very real possibility, even if that worst-case scenario is to run its course, there is time left for amends, honorable completions, and the chance to reconnect in to this Earth with the utmost respect, and in the gentlest of ways.[25]

Another example of a comprehensive and even more radical plan for adapting to the constraints demanded by the limits of the environment was developed in Switzerland in 1998. This intriguing plan had a double benefit: it addressed global economic inequality at the same time as it ameliorated the damages of the Anthropocene by the fossil fuel industry. The Board of the Swiss Federal Institutes of Technology proposed that all countries in the world reduce their overall energy consumption to 2000 watts per day per individual citizen by the year 2050. The 2000-Watt Society model can be replicated internationally by all societies and has as its central goal the downsizing and equalizing of economic activity and energy consumption globally. Their extensive research into global consumption of energy determined that 2000 watts was the average worldwide consumption. The 2000-watt limit would, of course, affect industrialized consumer societies most significantly. Europe, for example, would have to reduce its energy consumption by two-thirds; the United States by six-fold.

The 2000-Watt Society research studied many facets of the economy: cost reductions in heating, cooling, and constructing buildings; thermal power generation; all transportation systems – auto, truck, train, air; material use; among others.[26] While this model leaves the economic and political structures of capitalism in place, its strength lies in the overall concept of setting scientifically-determined objectives for downsizing – a bedrock of future earth-friendly sustainability.

An even more optimistic scenario for an economic transition has been proposed by journalist Paul Mason, author of *Postcapitalism*. He falls short of proposing an end to capitalism, but is aware of its shortcomings. He sees it transforming. He concluded that the current capitalist economic system which is based on industrial production is failing and will be replaced by an information driven system. This new economy, he argues, is not like current capitalism, and will generate a revolutionary new economic system grounded in an abundance of information. The transition is already underway: "New forms of ownership, new forms of lending, new legal contracts: a whole business subculture has emerged over the past 10 years, which the media has dubbed the "sharing economy". Buzzwords such as the "commons" and "peer-production" are thrown around, but few have bothered to ask what this development means for capitalism itself."[27]

While Mason continued to refer to the new system as a transitional phase within capitalism, his ideas merit some consideration precisely because capitalism and an emerging, more cooperative model can, theoretically, coexist during a period of transition. "If I am right, the logical focus for supporters of postcapitalism is to build alternatives within the system; to use governmental power in a radical and disruptive way; and to direct all actions towards the transition – not the defence of random elements of the old system. We have to learn what's urgent, and what's important, and that sometimes they do not coincide."[28]

Lewis and Maslin identified other optimistic options, expressing great faith in our collective human agency to problem-solve. They endorsed alternative energy sources – especially if they are decentralized; they predicted the potential for what they refer to as a new, "unmediated," as in free, information society; they proposed a universal basic income (UBI) to resolve economic inequalities; they made a passing reference to population control; and they embraced Edward Wilson's Half-Earth proposal to set aside half of the surface of the planet for other species. Left unsaid and unexplored by Lewis and Maslin are the current and future environmental costs of continued fossil fuel and/or alternative energy sources; the reality that all resources (natural and information) are highly controlled by capitalist enterprises; an uncritical endorsement of artificial intelligence (AI) and robots to create a potential utopian society; the question of pinpointing exactly which half of the earth would be set aside by whom; and how we would begin to limit our impact on the environment.[29]

Option Three
Live in an economy of precarity

Unfortunately, we are already well on the way to experiencing this less than satisfactory option. It is getting out of capitalism the hard way, as the economic system slowly disintegrates. Over the past seventy-five years, greed has accelerated global wealth inequality, forcing people worldwide to work two or more low-paying jobs just to provide minimal means of subsistence. People today are finding their economic systems already disintegrating, with a rise in homelessness, the loss of full-time predictable work and the shift to what cutely has been called the "gig" economy where workers compete for part time or contract jobs. This unstable economy has resulted in social unrest and growing frustration and anxiety.

An excellent analysis of a precarity economy is described in Anna Tsing's book, *The Mushroom at the End of the World: On the Possibility of Life in Capitalist Ruins*.[30] I had occasion to hear her deliver a keynote address at an international conference on Art and the Anthropocene, held in Dublin, Ireland in June 2019. In her comments she outlined how we might best respond to the changes we will be confronting. Among others, she suggested that we understand the effects of the Anthropocene as "patchy" both spatially and temporally; and that we should closely study the local impacts of the overall capitalist infrastructure on our lives to anticipate future problems and how we might respond to them. The fragmenting economy is not just about General Motors shutting down one plant. The more complex and far more devastating breakdown of many rural economies is already well underway. Those communities are microcosms of more widespread disruptions. The signs of coming decay are easily identified, and the challenges are on the horizon. Is our town dominated by one industry? What will we do if it suddenly closes? How will we get health care if the local hospital closes?

It will be especially important to understand something as simple as where food comes from and how its safety is ensured. Today, our food comes to us via a complex national and even international network. A couple of years ago, I stopped in the vegetable department of a large supermarket. Out of curiosity, I picked up a refrigerated package of peeled garlic buds. The garlic had been grown and processed in China. I had to ask myself, how is it profitable or even safe or wise to bring peeled bits of garlic across the Pacific Ocean to my supermarket in the middle of the United States? There is something deeply troubling about this model of food distribution, because garlic grows in my state of Minnesota and is widely available at farmer's markets. Tsing shared a similar example of how a global food economy can be especially detrimental. Blueberries picked in the fallout area from Chernobyl were shipped into the United States without disclosing their source, because of weak international agreements on agricultural products crossing multiple borders. If global food supply chains collapse, what will we do? Currently, California's Imperial County produces two-thirds of the vegetables consumed during the winter months in the United States. What if that ecosystem is threatened? Increasing soil salinity combined with the heavy dependence on irrigation will, over time, erode productivity in the Central Valley. Then what? Tsing also noted that we should pay attention to how the earth itself is used to meet the needs of capital rather than people – for example, mono-cropping and the energy intensive industrial farming methods that systematically destroy local biological systems. The message is that we all live downstream.

As humanity struggles to address the challenges of the Anthropocene, it will be a very stressful time politically. Some people – especially those with money – will be able to insulate themselves from the impacts and will find a niche where they can pretty much continue to go along and get along. Others will react with increasing anger, seeking to lay blame on any number of "others" whom they perceive to be the source of their precarity. Still others, out of feelings of increased anxiety, will be susceptible to authoritative leaders who will promise them stability. There is a real danger that some people will even welcome fascism and the promises of a powerful leader who will "save us" from ourselves.

Ecological crises resulting in economic instability are especially ripe for recuperation by fascist organizations. Fearful people are more susceptible to exploitation as they frantically seek to survive. Ecofascists are already merging environmental concerns about overpopulation with long-standing social divisiveness over immigration and multiculturalism, raising the specter of "lifeboat ethics" driving policy decisions. Ryhd Wildermuth, in an article entitled "The Future Is Fascist," argued: "Fascism – by which I also mean authoritarianism – is a way of managing civilizations during emergencies." He pointed to several global examples of rising repression: China's increasing surveillance system controlling access to resources via the social credit system; increasing taxes in France reflecting impending fuel shortages; and the ongoing migration crisis on the United States/Mexico border where the United States is building a wall. "The point of the wall isn't to keep people out now, it's to keep out the millions of people fleeing drought and starvation due to catastrophic climate change later. It is not about a racist present, but about a fascist future."[31]

The roots of today's ecofascism, according to environmental journalist Richard Smyth, can be traced to a small group of individuals who aligned the environmentalism of the Deep Ecology movement with white supremacy and even genocide, corrupting the philosophical concept of ecocentrism articulated by Aldo Leopold.

> From the misanthropic fringe of the 1980s Deep Ecology scene, for instance, Edward Abbey wrote of "culturally-morally-genetically impoverished" immigrants hampering his hopes for a "spacious, uncrowded and beautiful – yes, beautiful! – society" in the US. Abbey represented a vein of thinking in which the human species is considered at best to be an inconvenience and at worst a disease; take it far enough and it shades into Unabomber environmentalism. Ted Kaczynski, the American terrorist whose mail-bombing campaign from 1978

to 1995 killed three people and injured many others, talked a lot about nature – as a testing ground and shaping force, as a "counter-ideal to technology." In his justificatory "manifesto", he wrote: "The positive ideal that we propose is nature. That is, WILD nature: those aspects of the functioning of the Earth and its living things that are independent of human management and free of human interference and control.[32]

Obviously, I wish to emphasize and then underline, *not every environmentalist with an ecocentric philosophy embraces ecofascism*. Their earth-centered philosophy, however, has been co-opted, which unfortunately undermines the ecocentrist argument unjustly and inaccurately. To the contrary, paying attention to the needs of the planet logically requires paying attention to the needs of other human beings as well, reinforcing an ethic of environmental responsibility.

As we move forward, care must be taken to pay attention to the subtleties of various arguments and distinguish between a deep biocentric concern for the environment, a historical anthropocentric supremacy that privileges humankind over the environment, and the virulent anti-fascist, anti-humanity discourses that are aimed at protecting nature by attacking particular peoples or even humankind as a species. Discussions of overpopulation are more rationally grounded in concepts such as local environmental carrying capacity, drawdown, and depletion of natural resources rather than simplistic and dangerous charges of too many people – especially when there are references to too many of "those" people.

Option Four
Engineering the future – the technological fix

Human beings might choose to explore another option – doing nothing and trusting that future technological breakthroughs will save us in the end. New ideas are being floated daily, but to date they have been based on isolated experiments and have not been scaled up for global or even national implementation. A few examples being considered include: a new, non-polluting energy source that will emerge to allow us to leave the oil in the ground; a new bacteria that will disintegrate the plastics polluting the oceans without harming the sea beings; a new super being will be discovered that will digest all the excess carbon from the atmosphere; or that genetic engineering will protect us from the poisons we have created. Other proposals might embrace the idea that artificial intelligence (AI) machines will take over and manage our world, as explored by environmental philosopher Christopher Preston in "Forget the Anthropocene: We've Entered the Synthetic Age:"

> Unlike habitat destruction, carbon emissions, and other signatures of the Anthropocene epoch, the technologies being tested today are designed for consciously taking control of some of the key physical processes that shape our world. The bedrock laws of nature don't disappear, of course, but they become subject to a deeper kind of manipulation. You could think of these as not simply 'cosmetic' changes but 'metabolic' ones. Charles Darwin, Gregor Mendel, and the conventions of atmospheric physics become subject to a delicate kind of renegotiation.
>
> The crossing of this line represents radically new territory for both our species and for the planet. Nature itself will be shaped by processes redesigned and 'improved' by geneticists and engineers.[33]

David Wallace-Wells reflected on this topic in *The Uninhabitable Earth*, arguing that while we would like to believe that technology will save us, "more than tautologies" will be needed, "and, especially within the futurist fraternity of Silicon Valley, technologists have little more than fairy tales to offer."[34] Blocking the sunlight? How will that affect plant growth? Establish a colony on Mars? A more inhospitable climate could not be imagined, so why not just stay here and fix this one?

Alternative energy systems? On another technological front, we are currently accelerating the development and implementation of alternative energy sources around the world. We don't really know, and likely have not projected, the impact of these so-called "solutions" on how long the equipment will last, or on the true environmental costs of mineral extraction and industrial production of the needed materials, or the disposal of outdated equipment laced with toxic chemicals. Alternative energy sources come with environmental costs of their own – what Brett Bloom called "hidden externalities."[35] Further, Bloom argued that it only exacerbates a false optimism that things will remain the same: "One reason we think that wind power is an acceptable replacement for coal or oil is that we already accept that we are entitled to as much energy as we can use."[36] Wallace-Wells was not confident that we're making the right kind of progress. "Solar isn't eating away at fossil fuel use … even slowly; it's just buttressing it."[37] Technology will not necessarily save us:

> The scale of the technological transformation required … dwarfs every technological revolution ever engineered in human history, including electricity and telecommunications and even the invention of agriculture ten thousand years ago. It dwarfs them by definition, because it contains all of them – every single one

needs to be replaced at the root, since every single one breathes on carbon, like a ventilator.[38]

There is a lot of debate in the literature of the Anthropocene about geoengineering – especially tinkering with the atmosphere to cool the surface. This is a potentially dangerous option, as Jem Bendell noted: "The unpredictability of geoengineering the climate through the latter method [releasing chemicals to the upper atmosphere], in particular the dangers of disturbances to seasonal rains that billions of people rely on, make it unlikely to be used."[39] In the end, the dreams of technological solutions may be a fool's errand. For example, a recent Brazilian project released hundreds of thousands of genetically modified mosquitoes that were supposed to sterilize an indigenous mosquito population. Follow up studies showed that it didn't happen, and the law of unintended consequences was once again manifested. The modified mosquitoes are still breeding and spreading the new DNA, and no one knows how that will impact the "normal" population.[40]

Future technological production is also likely to be constrained by increasingly scarce resources. Already, significant amounts of fossil fuel must be used to extract oil from tar sands, and while fracking oil and especially natural gas is promoted as providing a low cost alternative to oil, the price of extraction includes the uncontrolled release of methane gas, the danger of earthquakes, and the polluting of the fresh water resources required to force the oil up to the surface. The resulting waste water is then generally pumped back into the ground further endangering fresh drinking water aquifers. There is a fancy term for this increasing cost of energy production – Energy Returned on Energy Expended (EREOI) as it gets harder to find and extract minerals and fossil fuels. Because capitalism privatizes profit and socializes the environmental costs, society and the environment will bear the costs down the road. In addition, many of the newer alternative energy technologies are heavily dependent upon non-renewable resources, as explained by writer Jasper Berns.

> That's because nearly every renewable energy source depends upon non-renewable and frequently hard-to-access minerals: solar panels use indium, turbines use neodymium, batteries use lithium, and all require kilotons of steel, tin, silver, and copper. The renewable-energy supply chain is a complicated hopscotch around the periodic table and around the world. To make a high-capacity solar panel, one might need copper (atomic number 29) from Chile, indium (49) from Australia, gallium (31) from China, and selenium (34) from Germany. Many of the most efficient, direct-drive wind turbines require a couple

pounds of the rare-earth metal neodymium, and there's 140 pounds of lithium in each Tesla.[41]

Think about this last bit of information when you feel like you absolutely must have a new cell phone!

The internet itself is a growing consumer of energy. In 2019, the Swedish KTH Royal Institute of Technology estimated that ten percent of the world's electrical consumption is dedicated to internet and associated information technology industry use, with projections up to twenty-one percent by 2030.[42]

There are also political ramifications to consider in evaluating the merits of any technological projects. Beginning in the nineteenth century at the onset of the Industrial Revolution, protests by the British weaver's guilds resulted in the loom-smashing movement called Luddites, after their leader Ned Lud. The Luddites recognized that technology was the lover of the Industrial Revolution and the benefits of the so-called "progress" of using machines rather the hand looms resulted in the loss of jobs and the centralization of power and wealth. Since that era, leftist critics – anarchists in particular – have been wary of technological advances, in part because the more complicated technologies become, the more they require a centralized command and control structure to maintain them.

Today we live within a centralized, authoritarian technology that still peddles the capitalist dream of progress and abundance. Technology controls every facet of our lives, making us ever more dependent, taking control through the universal obeisance to surveillance data, standardization, and systemization. Further, there is a sense of inevitability, which directly contradicts ideas of freedom and independent action.

So what are we to do if technological solutions are ineffective or unachievable? We have to assess, and then dismantle those technologies that threaten our continued freedom to live as a species and are destroying the planet, but we can't just start dismantling the machinery one factory at a time (although the system may already be crumbling). Dismantling technology will require a sophisticated plan. A classic case study of the failure to plan for technological change is what is currently happening in the declining coal industry in the United States. Rather than plan for shutting down the production and burning of coal and creating new jobs or educational opportunities for displaced workers, Donald Trump's solution was to remove pollution controls and allow the industry to continue dumping waste into local watersheds, destroying local environments.

As technologies fail, or are dismantled, people's lives will become even more precarious as jobs disappear. Many years ago, I visited the economically depressed Appalachian coal region. One evening, a group of us were sitting around a fire talking, when someone mentioned that a cousin had just gotten a job at a newly-opened hazardous waste dump. I commented that this toxic dump would pollute the area. His response was, "Lady, you don't live here, so don't tell us what to do. We need jobs." Without planning, disruptions could be profound.

Before relying strictly on technological or engineering fixes, we need to ask ourselves a series of questions. Are there good and bad technologies? What natural resources are consumed in production, and what is the state of nature afterward? Which technologies should we keep, which should we reject? Biomedical advances? The space program? Power plants? I-pods, radio, television? Public transportation? What about the internet? It requires a huge investment in infrastructure. What technologies are good for us? For the earth? Should we freeze all new technological developments while we sort this out? How much centralization is enough?

We can start with a grand conversation to assess and redefine the role of technology in our culture. Anarchist theory has consistently explored this theme, generating a wealth of ideas. What would a sustainable technology include? What could we do without? Anarchists have long made a strong case to return to decentralized, low (but efficient) technologies and need-based economic activities that could go a long way toward preserving the earth.

It is also time that we link political action to the issues raised by the Anthropocene, and assure that people have a voice. In his book *Nature, Technology and Society,* Mulford Q. Sibley, a political philosopher with socialist, pacifist, and anarchist tendencies, quoted a medieval proverb— *quod omnes tangit ab omnibus approbetur.* "Whatever touches all should be approved by all." He suggested two principles to guide re-evaluating any technology: the burden of proof should be on the creators of technology to prove that it will do no harm, and perhaps even be of supreme benefit; and there should be a democratic up or down vote about whether the technology should be adopted or dismantled.[43]

Option Five
The unthinkable

An even more perilous and unpredictable option is that we do nothing and nature just takes its course, and we end up with "mutually assured destruction" – a phrase coined during the Cold War to describe the potential for a catastrophic nuclear war with Russia. Philosopher Jason Read argued that capitalism may be undone in the end by nature anyway, rather than anything human beings could do. The real battle takes place on a higher level with higher stakes – capitalism versus the hyperobject of global warming. Nature, resisting exploitation, may simply undo aspects of capitalism in acts of self-preservation:

If the Anthropocene can in part be understood as a kind of multiplication of natural limits, as oceans, the atmosphere, and various ecosystems reveal their irreducibility to abstract nature, to nature for capital, the anthropogenesis can in part explain why resistance to capital is found more on the side of nature than culture. Nature is surprisingly resistant, and we ourselves are incredibly inert. Or, more to the point, some aspects of nature are rebellious, constantly rebelling against its status as cheap nature, imposing the costs of ecological collapse, antibiotic-resistant bacteria, and desertification.[44]

Option Six
Living outside of capitalism, building a parallel, earth-friendly world.

Most of the above options implicitly assume various re-configurations of centralized government decision-making and the maintenance of the capitalist economic system. An alternative option, which I will explore in the following chapter, is adopting the principles and practices of social anarchism/libertarian socialism. This option provides an opportunity to introduce new political goals, practices, and ethical approaches, and to generate more creative options as the petro-world crumbles around us. The good news is that there is no need to wait for the collapse to implement many of the alternative actions.

NOTHING WILL EVER BE THE SAME AGAIN
ANARCHIST PRACTICE FOR THE ANTHROPOCENE

> Think about the world you want to live and work
> in. What do you need to know to build the world?
> Demand that your teachers teach you that.
> – Peter Kropotkin

THIS CHAPTER BEGINS WITH A DEPRESSING ASSUMPTION: IF NOTHING IS DONE soon by national and global leadership to begin the process of dramatically reducing fossil fuel use and curtailing rampant consumerism, there will be an accelerating period of economic and political collapse. Climate changes will occur unevenly around the globe: fires will affect some, rising sea levels will inundate coastal communities, weather will become increasingly violent, food insecurity will increase, and declining natural resources will result in gaps in building and maintaining current infrastructure.

Environmental and cultural instability will result in increasing individual and cultural stressors creating anxiety and unease, the possibility of increasing violence, and the collapse of social networks. These breakdowns will first appear in communities and countries living in ecologically sensitive areas, then cascade randomly through societies worldwide. Human tragedies will escalate in already economically precarious environments and especially in those populations forced to migrate.

We could all stand by and watch this slow motion environmental train wreck. Or we can begin to develop and practice environmentally aware ways of thinking and living right now, while we still have time. We can start by asking ourselves a few questions.

- How can we change individual behaviors – all with an eye to reducing adding CO_2 to the environment?
- How can we identify and establish the reduced limits within which we can live?
- How well do we understand the natural networks in which we live?
- How well do we understand the effects of current economic systems and the fossil fuel economy on our daily lives?
- How are we networked with other like-minded people – and with whom could we work/find/share skills for living more collectively?
- How will we identify the common good and work towards achieving it with other people?

According to Senior Fellow at the Brookings Institute Elaine Kamarck, there are four reasons why people have trouble accepting the realities of global warming and then acting politically: its complexity; the appropriate jurisdiction and accountability for addressing it; the challenge of building collective action and trust; and nourishing an imaginative spirit.[1] The complexity of global warming requires two seemingly contradictory responses: an appreciation for the global nature of the problem, i.e., it's not just my town or my country; and the feeling that the problem may be too large to address at all. Further, no formal global structure has yet been recognized that can articulate for the general public the complex scientific nature of global warming, identify the sources of problems and hold individual countries accountable, or effectively coordinate responses for mitigation. How will we decide whether national or global limits should be set, where taxes should be assessed, or how cooperative enforcement agencies can be established? Building planet-wide collection action and trust will be a big hurtle to overcome.

Once responsibilities are identified and accountability mechanisms agreed to, the next major challenge is building collective consensus around mitigation and its enforcement. Who will be the first to turn off their air conditioning? Which government sets the standards? Which nation reduces its fossil fuel consumption to accommodate growing economies? Will we vote for electricity bills that reflect penalties for exceeding limits? Without trust in the ability to work with others, mitigation will remain only a fantasy.

Think of the many ways we already cooperate that could provide models for developing a collaborative society. Those insights, social practices, and skills can become a global new norm. It can be done. For example, when we eat together and pass the bowl of potatoes around the dinner table, everyone takes some, always thinking of leaving some for the next person. Our challenge will be to establish similar patterns of cooperative relationships with special attention and engagement with the other beings with whom we share the planet. Biologist and sustainable environmental designer Daniel Wahl calls this process of collaboration "interbeing" – the understanding that we live in a cooperative, mutually-sustaining environment.[2]

An additional benefit of resource sharing, of course, is built-in equitability. All members of the ecosystem will be guaranteed equal representation and access to resources. Once resources are identified and catalogued, and ecosystem limitations have been identified, collective decision-making structure can ensure that those resources are distributed in a manner that reflects the needs of all members of the community, including those non-human members in the immediate environment.

A Quick Introduction to Anarchism

Anarchists have an important role to play in addressing the emerging ecological crisis. An uncertain and constantly-changing world will need new, bold ideas, what Kamarck calls an "imaginative spirit." Anarchist theory is uniquely equipped to point out the enemies of life, address changing expectations, and deal proactively with uncertainty in a time when people are seeking a vision, a hope, a different dream that is realizable. Who else has a Big Vision that isn't tied to money and power dreams and is committed to engaging the critical players in a collaborative process as we move forward? What other theoretical model is open and flexible enough to address and creatively respond to the different and changing challenges confronting us?

In the liminality of an environmental apocalypse, social anarchism can provide a hopeful way forward through the uncertainty. The basic tenets of the political theory of social anarchism or libertarian socialism, are outlined below by philosopher Nathan Jun:

> (1) unqualified moral and political opposition to relationships and institutions based on coercion, domination, oppression, and other forms of arbitrary and unjustifiable authority; (2) unqualified moral and political opposition to all forms of arbitrary and unjustifiable inequality (whether political, social, economic,

sexual, or cultural); (3) an active commitment to combating and ultimately eradicating unjustifiable authority and inequality by means of organized direct action; and (4) an active commitment to building non-authoritarian, non-hierarchical, and egalitarian relationships and institutions based on communal self-management, voluntary association, and mutual aid.[3]

Anarchism is not a rigidly proscribed political ideology, nothing is set out as a utopian process to be achieved in a measured linear way. It is rather a political activism embodying a practice constantly renewing itself as it responds to a dynamic environment. It reflects sensibilities attuned to the multiple contexts – environment, connections, relationships, interactions, and realistic solutions – by providing a model for collective action. A key feature of anarchist practice is the freedom and flexibility to respond to changes in the immediate reality, adapting actions on the fly. Moving in a more environmentally sustaining lifestyle means continually reassessing and recreating new solutions – an important skill in an uncertain environment. In discussions with artist Gee Vaucher, Stevphen Shukaitis reflected on the importance of openness and adaptability to respond effectively to emerging environmental crises: "It's particularly interesting that Gee has suggested that if she has a relationship to anarchism it's in always throwing her methods and assumptions into question. She never wants to be stuck in something; she always wants to rethink how she can do things differently."[4] This is the perfect politics for a liminal time and space.

So how to begin?

The collapse of the complex social, political, and economic structures of contemporary capitalism, will require the skills and commitment of a population open to change, to thinking on their feet, and interested in working with others, including with the earth. Anarchists are not afraid of trying new ways to organize communities and are not interested in ceding decision-making to a distant power structure that is not nimble enough to coordinate nearby collapse. We need only to remind ourselves of the United States government's appallingly inadequate response to hurricane Katrina. In the ensuing chaos, the anarchist Common Ground Collective was one of the first groups to provide aid to the residents of the hard-hit New Orleans Ninth Ward.[5] Focusing on decentralized, action-oriented anarchist political practices will provide an opportunity to more quickly identify the problems and make conscious, proactive responses in a timely manner. The closer you are to a problem, the quicker you can respond – an obvious benefit of decentralized political organizations.

A Brief Overview of Anarchist Environmentalism

Anarchists need a critical environmental practice to comprehensively address the coming social and economic upheaval. The most profound failure of the environmental movement for the past fifty years is this: the majority of mainstream environmental organizations consciously choose to work within federal and state legal and regulatory mechanisms to advance a conservation agenda. To their credit, they were able to preserve and protect many environmentally sensitive areas, but they failed to address the underlying problem – an entire political and economic system built on environmental exploitation – and they never sought an alternative political method to critique this system and develop alternatives. Corporate capitalism and its commitment to ongoing destruction of natural resources and the fantasy that modern science and more advanced technology would eventually "fix" all problems have only served to fuel denial and resistance to addressing the consequences of global warming.

Another philosophical and tactical error of early environmental groups was the decision to adopt a narrow, individualistic, consumerist approach to addressing global warming. Everyone was told they had an individual "carbon footprint," which they were individually responsible to reduce. In this approach, the economy will continue to hum along and you can keep doing what you're doing, but plant a few trees, or better yet, if you have money, just "buy" your piece of guilt-free carbon sequestration. But we are not going to shop our way out of an environmental meltdown by putting a price on carbon emissions and selling them globally. An individualistic approach has obscured the collective nature the destruction and the necessity for collective action to address the damages and establish new practices that are environmentally neutral. While we might feel good about taking our beer cans down to the recycling center, we are not going to recycle our way out of an environmental meltdown. Do we really believe that the planet can be saved one random act of rejecting single use plastic bags at a time? Or that we can advertise our way out of this crisis?

An even more serious consequence of making individuals responsible for mitigation meant that people who adopted helpful actions might feel free to wash their hands of any collective responsibilities. The individual is still a member of the species, the "we" who are damaging the planet. There is no escaping the network of our personal and social responsibility.

What is needed instead, and what social anarchism can provide, is a total radical rethinking of how we can collectively respond to restructuring the current economic and political systems. French author Herve Kempf observed that "it is not enough for society to become aware of the urgency of the ecological crisis. … It will further be necessary that ecological concerns articulate themselves as a radical political analysis of current relationships of domination.

We will not be able to decrease global material consumption if the powerful are not brought down and if inequality is not combated. To the ecological principle ... 'Think globally; act locally,' – we must add the principle ... 'consume less; share better.'"[6]

The scientific study of the human impact on the environment has a long history. From the sixteenth century on, global exploration brought new and exotic animal and plant life to the attention of early European scientists and collectors. The Enlightenment, with its discoveries and celebration of science, privileged an empirical scientific method and the critical examination of the planet and the heavens.

By the nineteenth century, scientists and philosophers like Vasily Dokuchaev (the father of soil science) in Russia, Charles Darwin (evolution) in England, Henry David Thoreau (observations at Walden Pond) in the United States, and others too many to mention were exploring, dissecting, and cataloging the workings of the environment. These early scientists were also aware of how humankind had negatively already impacted the environment. Soils were wearing out, the smoke pollution from the burning of coal darkened the skies over urban centers, and runoff from factories polluted rivers and drinking water. As knowledge of the natural world grew, concerns about the impact of industrialization also expanded. Political economists Karl Marx and Friedrich Engels, for example drew parallels between the exploitation of workers and the exploitation of natural resources. They also recognized the interdependence between human economics and the natural world. In the *Economic and Philosophical Manuscripts of 1844* Marx wrote: "In a physical sense, man lives only from these natural products, whether in the form of nourishment, heating, clothing, shelter, etc. ... Man lives from nature – i.e., nature is his body – and he must maintain a continuing dialogue with it if he is not to die. To say that man's physical and mental life is linked to nature simply means that nature is linked to itself, for man is a part of nature."[7] In other works Marx studied soil depletion and the air pollution from industrialization. In "The Part Played by Labour in the Transition from Ape to Man," Friedrich Engels shared a keen awareness of the negative impact of human economic activity on the environment, even as he continued to accept the Enlightenment perspective of the supremacy of human rationality in justifying control of the natural world.

> Let us not, however, flatter ourselves overmuch on account of our human victories over nature. For each such victory nature takes its revenge on us. Each victory, it is true, in the first place brings about the results we expected, but in the second and third places it has quite different, unforeseen effects which only too often cancel the first. ... When the Italians of the Alps used up the pine forests on the southern slopes, so carefully

cherished on the northern slopes, they had no inkling that by doing so they were cutting at the roots of the dairy industry in their region; they had still less inkling that they were thereby depriving their mountain springs of water for the greater part of the year, and making it possible for them to pour still more furious torrents on the plains during the rainy seasons. ... Thus at every step we are reminded that we by no means rule over nature like a conqueror over a foreign people, like someone standing outside nature – but that we, with flesh, blood and brain, belong to nature, and exist in its midst, and that all our mastery of it consists in the fact that we have the advantage over all other creatures of being able to learn its laws and apply them correctly.[8]

Classical anarchists – many of whom were familiar with Marx's writings – began raising similar concerns. According to Matthew Hall, nineteenth century anarchists were mixed in their attitudes toward nature. He argued that while we can be critical of their human focus – they were anthropocentric for the most part – their thinking was likely the legacy of their Enlightenment roots. Nevertheless, nature was always part of early anarchist political and economic thinking.[9]

Perhaps the most influential of the late nineteenth century anarchists writing about the environment in the nineteenth century was Élisée Reclus, a French geologist and geographer. His writings on the environment that integrated his politics, ethics and nature studies, influenced subsequent earth scientists and anarchists. His political philosophy included a strong ethical component that translated to the necessity of caring for the earth. He was a prolific writer – his *The Earth and its Inhabitants: The Universal Geography* was a nineteen volume study of the environment and peoples of every continent.

Reclus tempered anthropocentric approaches with an appreciation of the importance of nature in the human experience. His philosophy centered on ideas of harmony and balance between nature and humankind, and is consistent with contemporary ecocentric approaches. In an excellent overview of Reclus's ideas, philosopher John Clark described Reclus's account of the dialectical and cooperative interaction between humankind and the planet.

> Reclus exhibits in all his works a strong sense of humanity's embeddedness in nature. ... Throughout his works, he continues this holistic, integrative approach. While his studies of the natural world became increasingly scientific and empirical, he never abandoned his early romanticist, poetic, and even spiritual attitude toward nature. Indeed, his quest to integrate forms

of rationality and imagination that have often been opposed to one another is one of the most noteworthy dimensions of his thought.[10]

Reclus emphasizes the need for a greater recognition of nature, not only in the sense of understanding its activity, but also in the sense of developing a new responsibility toward it. This concern underlies the scathing critique of humanity's abuse of the earth that he began to develop early in his work.[11]

Human progress and the very survival of the human species, according to Reclus, was based on mutual respect and caring for the environment. The role of human beings was to find their place, fit in, and act responsibly: "Having become the 'consciousness of the earth' the man worthy of such a mission assumes, by virtue of that, a responsible role within the harmony and beauty of his natural environment."[12]

> Human developments are linked in the most intimate manner to the natural environment. An implicit harmony exists between the earth and the people it nourishes, and when imprudent societies strike a blow against what beautifies their environment they have always ended in regretting it. ... Among the causes in human history which have already contributed to the disappearance of many successive civilisations, one must mention the brutal violence with which the majority of nations have treated the nourishing earth. They cut down forests, dried up springs, flooded rivers, damaged climates, surrounded the cities with swampy and pestilential zones, then, when nature desecrated by them has become hostile, they grasped her with hatred and not being able to re-immerse themselves like savages [sic] into the life of the forests, they let themselves become more and more stupefied by the despotism of priests and kings.[13]

Care of the earth included the recognition that there were other beings to be considered as well: "An area in which Reclus was far in advance of his time, and in which he anticipated current debates in ecophilosophy and environmental ethics, is in his concern with ethical and ecological issues regarding our treatment of other species."[14]

Early Russian anarchist philosopher, Peter Kropotkin, also trained as a geologist, and like Reclus, he had a keen interest in both anarchism and the environment and their impact on politics. Political philosopher Ruth Kinna linked Kropotkin's political theories on evolution and cooperation to his understanding of science, and his concern that over the span of the nineteenth century, humanity had moved away from nature. His best known work in political theory, *Mutual Aid: A Factor of Evolution,* was published in 1902, and was dedicated to the belief that human society was based on cooperative principles learned from his observations of the natural world. Kropotkin was also well-versed in the evolutionary model proposed by Charles Darwin, and *Mutual Aid* challenged Darwin's assumptions of competitiveness as the driver of evolution. While Kropotkin did not directly address environmental problems in the same way we do today, he grounded his political theory of cooperation and mutual aid in his scientific phenomena, focusing on how species interacted to enhance mutual biological survival. His observations of ants, termites, bees, eagles, wolves, and the primates reinforced again and again, the cooperative nature of a wide range of species. Hall regarded Kropotkin as a key contributor to an early anarchist theory that was also based on his scientific study of the natural world. Kropotkin was especially struck by the cooperative nature of animal society.

> [Kropotkin] alludes to the existence of society in the animal kingdom, a social organisation that is characterised more by mutual aid and reciprocity than by Darwinian struggle and competition. As he based his idea of an anarchist social organisation on these principles, it can be argued that Kropotkin greatly valued these non-human societies. Drawing direct inspiration from the animal kingdom also positioned humanity as another animal species, as a part of nature.[15]

Kropotkin focused his observations on species rather than individual behavior, arguing that there was little competition between individuals *within* species where cooperation was the rule. Instead, it was species as a whole that competed for survival against other species and the broader the forces of nature. "Kropotkin's most famous book, *Mutual Aid*, maintains that cooperation within a species has been an historical factor in the development of social institutions, and in fact, that the avoidance of competition greatly increases the chances of survival and raises the quality of life."[16]

It would not be until the middle of the twentieth century that radical environmental theories reached their full expression in the United States in response to the rising concerns about environmental damage generally. Anarchists played an important role in the emerging field of environmental justice, ethics and activism. While conservation efforts continued, attention turned to the growing concerns about such environmental issues as industrial and agricultural pollution, governmental mismanagement of public lands, overpopulation, and rampant consumerism.

The philosophical positions of these mid-century environmentalists ranged between two perspectives – anthropocentrism and ecocentrism. Both groups were concerned about the emerging crises, but developed their responses depending on their different beliefs. The anthropocentrists prioritized the needs and wants of humanity. For example, their efforts were concentrated on setting aside wilderness areas for human enjoyment or calling for governmental regulation of pollution. At the same time, and with the same logic, they looked the other way when western grasslands were overgrazed by beef cattle ranchers' exploitive practices, and mining and timber cutting were allowed on other public lands to fuel the ever-expanding economy. The ecocentrists, on the other hand, were adamant from the beginning that the needs of nature came first, and the correct political stance was an activist, radical defense of the biosphere as a whole by supporting its diversity, complexity, and fighting for the survival of all species. Anarchists for the most part chose the latter path, advocating for direct action against human activity on behalf of the planet. The government could not be trusted to protect the environment.

The Deep Ecology movement reflected this shift to a more ecocentric philosophy that focused theory and debate on the needs of nature. Based on the writings of Henry David Thoreau, Lev Tolstoy, and John Muir, and especially Norwegian philosopher Arne Naess, the term Deep Ecology described a biocentric and ecocentric environment characterized by reciprocity between humanity and nature: "The ecosophical outlook is developed through an identification so deep that one's own self is no longer adequately delimited by the personal ego or the organism. One experiences oneself to be a genuine part of all life. We are not outside the rest of nature and therefore cannot do with it as we please without changing ourselves."[17] Along the same lines, British chemist James Lovelock formulated the Gaia Principle which argued that the planet was an integrated system where organic and inorganic beings lived in dynamic balance to ensure life for all. His theories were consistent with those insights developed by Russian scientist Vladimir Vernadsky's into the biosphere and Peter Kropotkin's theory of mutual aid.

While he did not identify as an anarchist, E. F. Schumacher was another important environmental voice in the 1970s. He captivated an early generation of environmentalists and anarchists with his book, *Small Is Beautiful: Economics As If People Mattered*. He proposed a Buddhist economic model, a "Middle Way" in all things, advocating the importance of appropriate scale, with a preference for a local and decentralized context for most economic and political activities; working in a collaborative manner with the earth and husbanding the use of resources; and designing appropriate local, small scale technologies. Regarding an appropriate political scale, Schumacher concluded that local and global perspectives were equally necessary: "When it comes to action, we obviously need small units, because action is a highly personal affair, and one cannot be in touch with more than a very limited number of persons at any one time. But when it comes to the world of ideas, to principles or to ethics, to the indivisibility of peace and also of ecology, we need to recognize the unity of mankind and base our actions upon this recognition."[18] Decentralized, local social and political units are a hallmark of anarchist thinking – a philosophy of parsimony that Schumacher argued, is consistent with living close to the earth and to the materials of making a balanced life.

In a direct rebuke of consumer societies, the nature of appropriate resource use was outlined in the chapter on "Buddhist Economics." According to Schumacher, "Non-renewable goods must be used only if they are indispensable, and then only with the greatest care and the most meticulous concern for conservation" – economic ideas that challenge the foundations of advanced capitalism and are consistent with an ecological sensitivity.[19] Environmental historian Keith Woodhouse chose the term "forebearance" to describe this concept of learning to living wisely within nature's parameters – "humility, precaution, and the inclusion of nonhuman interests in human decision-making."[20]

The concept of small is beautiful raised the question of the carrying capacity of local and global ecologies, especially as human populations grew and placed increased pressure on the livelihoods and survivability of other, usually wild, plant and animal beings. Scale and carrying capacity are important considerations – for the planet, for food production, for livable cities, for organizing collective politics. "For radicals, the central concern of environmentalism was always with limits: to natural resources, to industrial expansion, and to human population."[21] In books such as Paul Ehrlich's *The Population Bomb;* the Club of Rome's report *The Limits to Growth;* Herman Daly's *Steady State Economy;* and Schumacher's *Small is Beautiful* the outlines of the challenges to be confronted were carefully laid out. The rhetoric, however, continued to focus arguments on humans, not nature, according to Woodhouse.[22]

Mid-twentieth century anarchist philosophical and political positions in the United States dovetailed nicely with a burgeoning radical environmental movement emphasizing an ecocentrist philosophy. Environmental activism shifted from conservation and wilderness preservation for human enjoyment to what Woodhouse called "crisis environmentalism." By the late 1970s, Earth First! emerged as the preeminent radical environmentalist organization practicing an ecocentric philosophy. It was founded by a core of radicals who had broken away from more conservation-oriented mainstream organizations such as the Sierra Club and the Wilderness Society, to take up an activist role as defenders of the biosphere as a whole. Deep Ecology principles, coupled with the anarchist practice of direct action, were adopted and practiced by radical groups like Green Peace, Earth First! and the Earth Liberation Front. Their philosophy and direct action praxis was, and continues to be replicated globally to this day. These activist ecocentric organizations moved the environmental debate beyond the conservationist, anthropocentric policies and practices of the early environmental movement.

While Earth First! did not abandon a previous commitment to preserving wilderness areas, they did reject mainstream environmental organizational tactics of negotiating over governmental regulatory procedures and advocating legal protections. Instead, they turned to anarchist tactics of direct action and even sabotage in active defense of the environment. They were uncompromising in their criticism of expansionist capitalism and according to Woodhouse, "they cast a wary eye on much that humans thought and did, and that wariness guided their politics."[23] Earth First! and Edward Abbey made major contributions to the radical environmental movement, including bringing attention to direct action as a tactic, raising awareness of how we were destroying the environment; and their appeal to American romanticism – the freewheeling cowboys on behalf of the environment. They adopted the anarchist principle of direct action – practicing civil disobedience, "monkey-wrenching" to disable forestry and mining equipment, long-term occupation of threatened ecologies, tree-sitting to save old growth forests, bringing down powerlines, and cleverly detourning or even dismantling of public advertisements: "The Ann Arbor *Argus* told of a group of 'billboard bandits' who targeted roadside advertising throughout Michigan."[24] The tactic of changing messaging in public spaces was widely adopted by many groups on the Left. [This author recalled a personally satisfying experience spray-painting and "revising" the sexist message on a three-story billboard!] Earth First! inspired other radical environmentalist activities including the journal *Live Wild or Die*, and the pamphlet "Alien-Nation" written by the West Coast group Washington Earth First! The *Earth First! Journal* is still published bi-monthly today, and is a rich source of information on radical environmental activism worldwide.

There was some divergence of opinions between radical environmentalists and left political activists. Charges were volleyed black and forth as environmentalists chided Leftists for ignoring environmental issues, and Leftists charged that the environmentalists had only a tepid critique of capitalism, and tended to ignore social and economic issues affecting people. The radical environmentalists always came down on the side of nature. According to Woodhouse, however, environmentalists were well aware of the political positions of the prevailing political and economic power structures: "Radical environmentalists shared anarchists' dim view of government; … the complexity of modern technology; the compromises and corruption of representative democracy; the misguided emphasis on the individual by liberalism; and the exploitative and utilitarian use of the natural world by industrial society."[25] Their philosophies continued to clash over whether nature or humanity took precedence when the two perspectives differed.

Murray Bookchin attempted to bridge the theoretical gap between the broader Left and environmentalists by merging the two political agendas in his theory of social ecology: "Social ecology argued, essentially, that people's abuse of the natural world resulted directly from social inequality, that control and exploitation among human beings of each other led to control and exploitation by human beings of nature."[26] Bookchin was especially critical of Deep Ecology and radical environmentalists like Earth First! and anti-civilization anarcho-primitivists like John Zerzan.

Bookchin was deeply committed to local, decentralized political organization and action. According to anthropologist Katie Horvath,

> The beauty of dual power as a framework is that we can make concrete improvements to our daily lives while simultaneously laying the groundwork to challenge capital and the state. By constructing grassroots, horizontal, local institutions that take the place of exploitative or absent statist and/or capitalist institutions in our and our neighbors' lives, we are both planning for the future and meeting our needs right now.
>
> Hierarchical institutions are inherently unaccountable to the people most affected by the climate crisis; we cannot expect the institutions responsible for the crisis to be capable of managing our collective future going forward. At best they will buy us a couple of decades.[27]

An even more extreme environmental theory, Primitivism, took ecocentrism to one of several logical conclusions. Primitivists argued that humankind was such a scourge on the planet that human activity had to be dramatically curtailed. Primitivism's principal proponent was anarchist John Zerzan. He coined the term "future primitive," to describe the political objective of decentering the role of humanity as a superior force in nature by dismantling civilization and returning humankind to its primitive, hunting gathering level. His radical ecocentric theory celebrated the deep connection between all life forms. He based his thoughts in part on his readings of Theodore Adorno's analysis of the failures of civilization combined with the political practices embedded in anarchism. Anarcho-primitivism advocated "rewilding," de-industrialization, and the breakup of complex social organization. The contemporary primitivist Green Anarchy Collective summarized Primitivism as a sustainable rewilding process for both humans and the earth:

> For most green/anti-civilization/primitivist anarchists, re-wilding and reconnecting with the earth is a life project. It is not limited to intellectual comprehension or the practice of

primitive skills, but instead, it is a deep understanding of the pervasive ways in which we are domesticated, fractured, and dislocated from ourselves, each other, and the world, and the enormous and daily undertaking to be whole again. Rewilding has a physical component which involves reclaiming skills and developing methods for a sustainable co-existence, including how to feed, shelter, and heal ourselves with the plants, animals, and materials occurring naturally in our bioregion. It also includes the dismantling of the physical manifestations, apparatus, and infrastructure of civilization.

Rewilding has an emotional component, which involves healing ourselves and each other from the 10,000 year-old wounds which run deep, learning how to live together in non-hierarchical and non-oppressive communities, and de-constructing the domesticating mindset in our social patterns.[28]

Primitivist ideas never gained purchase, perhaps because of the horrifying and almost incomprehensible implications of dismantling our contemporary civilization. The movement has also suffered from a negative public image – in part, because primitivist theories were propagated by the Unabomber Ted Kaczynski and because they have recently found a receptive audience with the far right and ecofascists. Timothy Morton reinforced the manner in which fascism and racism converge and threaten environmental politics:

The Left should take heed that the Far Right underpins speciesism with racism by fusing paranoia about biodiversity with anti-Semitism. The struggle against racism thus becomes a battleground for ecological politics. "Environmental racism" isn't just a tactic of distributing harm via slow violence against the poor. Environmentalism as such can coincide with racism, when it distinguishes rigidly between the human and the nonhuman. Thinking humankind in a non-anthropocentric way requires thinking humankind in an anti-racist way.[29]

The tendency to see nonhumans as unthinking and even unfeeling machines is predicated on the objectification and dehumanization of other humans, not the other way around.[30]

A more recent and growing trend in global environmental politics, is the international Rights of Nature movement. The Nature Rights movement advocates are ecocentric in the sense the claim to a "right" is based on recognizing the agency of nature as a political co-participant. Implicit in the Rights agenda is a repudiation of the capitalist principles of property ownership and exploitation. Nature ceases being property and becomes Being with a right to exist. As the Global Alliance for the Rights of Nature explains on its website:

> Under the current system of law in almost every country, nature is considered to be property, a treatment which confers upon the property owner the right to destroy ecosystems and nature on that property. When we talk about the 'rights of nature,' it means recognizing that ecosystems and natural communities are not merely property that can be owned, but are entities that have an independent *right to exist and flourish*. Laws recognizing the rights of nature thus change the status of natural communities and ecosystems to being recognized as *rights-bearing entities* with rights that can be enforced by people, governments, and communities.[31]

Although not a specifically anarchist project, the idea of natural rights demanding legal standing for the natural world has attracted interest by some national governments and indigenous tribal governments. There are efforts underway, spearheaded by Bolivia, to seek United Nations endorsement for these principles.

To date, Rights of Nature have been granted by several countries, including Ecuador who embedded the rights in their Constitution, and Bolivia passed the Law of Mother Earth in 2010. New Zealand granted rights to the Whanganui River in 2017. In the United States, Minnesota Anishinaabe spokesperson and environmentalist Winona LaDuke explained that the White Earth Band, has granted rights to other beings, including plant relatives. These rights are guaranteed in Anishinaabe law. The Rights of Manoomin, for example, state that "*Manoomin, or wild rice, within all the Chippewa ceded territories, possesses inherent rights to exist, flourish, regenerate, and evolve, as well as inherent rights to restoration, recovery, and preservation.*" The Rights include: "The right to clean water and freshwater habitat, the right to a natural environment free from industrial pollution, the right to a healthy, stable climate free from human-caused climate change impacts, the right to be free from patenting, the right to be free from contamination by genetically engineered organisms."[32]

The Rights of Manoomin were modeled after the Rights of Nature, recognized in courts and adopted internationally. To date, several other tribal governments have adopted the Rights of Nature. The Yurok Tribe in California granted protected rights to the Klamath River; the Ho-Chunk Nation of Wisconsin moved to protect the earth on their nation from fracking and the poisonous practices of industrial agriculture, and the Ponca Nation in Oklahoma protected fish, and rejected fracking.[33]

In 2019, voters in the city of Toledo, Ohio approved the Lake Erie Bill of Rights that includes the following statement:

> And since all power of governance is inherent in the people, we, the people of the City of Toledo, declare and enact this Lake Erie Bill of Rights, which establishes irrevocable rights for the Lake Erie Ecosystem to exist, flourish and naturally evolve, a right to a healthy environment for the residents of Toledo, and which elevates the rights of the community and its natural environment over powers claimed by certain corporations.[34]

The Bill of Rights was summarily killed by the Chamber of Commerce with the passage of a state law prohibiting suing polluters.[35]

It goes without saying that rights for nature will benefit humankind. We are, after all, nature, too.

Tools for a new world

Changing the future begins with resistance to the status quo. The sirens of corporate capitalism are already at work soothing fears and lulling us into believing that the "free market" can best solve environmental problems painlessly, and that even big corporations can become green overnight. This position is painfully obvious when we examine corporate dialogue about the fossil fuel industry. Exxon is already actively promoting carbon capture and storage (CCS) technology as a cheery solution to rising carbon dioxide increases caused by the burning of fossil fuels. What their slick television and print advertising doesn't tell us is that the technology is costly; that using more fossil fuel energy is required to pull carbon out the atmosphere; and a couple hundred thousand facilities will be needed to lower levels – just in the United States. Only a few prototypes have been built, but these messages to the public are designed to soothe concerns and smooth over the problems with optimistic fantasies. We are fed a constant stream of consoling talk by the energy producing industry, assuring us that alternative energy sources will reduce carbon in the atmosphere. Natural gas has been heavily promoted as reducing carbon in the atmosphere, eliding the fact that it, too, is a fossil fuel and thus not really solving the problem. In addition, its extraction is environmentally costly.

At the same time the energy consuming industry continues to expand apace, essentially cancelling out any reduction in carbon output. Automotive manufacturers continue building (and promoting to eager consumers) larger sport utility vehicles (SUVs) and pickup trucks. Since 2010, in the US, 35 million more of these gas-guzzling vehicles are on American roads, and production of small, efficient vehicles has virtually halted. According to Nathan

Johnson, these SUVs, are "warming our planet more than heavy industry."[36] Over the same time period, only five million electric cars were produced, but as more are produced, fossil fuels will still be used to produce the electricity to power them.

Many governmental and corporate "solutions" are focused on individual actions, rather than restructuring industrial and energy systems. We are advised to take up recycling, drive an electric car, use an Uber or Lyft car instead of driving ourselves, take fewer showers, and install home solar panels. These are nice to do, but are, in the long run, ineffective and unrealistic options. One of the central precepts of the neo-liberal agenda is the idea of the individual as free agent operating in a free, capitalist market, and every effort is made to maintain this illusion of continually expanding options and choices. The entire capitalist economy and its advertising arm encourage continuing consumption, focusing on individual desires rather than offering real collective solutions to real problems. As Amitav Ghosh noted, "global warming poses a powerful challenge to the idea that the free pursuit of individual interests always leads to the general good."[37]

Even as awareness of the climate crisis grows, the market continues defining for us our collective needs and transferring those desires into a consumer culture driven by the belief in personal choice. Never do we hear that we should consume less, or not drive at all. We never stop to think that the ride sharing services Uber, Lyft and self-driving cars are merely continuations of the cult of the car, and are now clogging city streets with even more traffic. If you or I drive to the store in our personal car, we make two trips, to and from. A ride share vehicle, on the other hand, makes four trips: two to pick you or I up, and two to complete the trip to the store. The whole convenience car industry is built on a "gig" economy model, extracting profits from desperate individuals who are basically underwriting the whole enterprise with the purchase and maintenance of their personal automobiles. The ethical choices are simple: do we accept these cars trolling the streets endlessly for fares (and all the while consuming fossil fuels) or do we discourage individual driving and support public transportation? I'm still waiting to see some in depth analysis, for example, on whether shopping from home (and even worse, the practice of next day delivery on demand) is really better for the environment. It is true that businesses can save significant dollars in rent and employee wages by closing walk-in shops, but no one so far has analyzed the fossil fuel and street maintenance costs of all those delivery trucks on the road. Seldom are these individual actions evaluated for their "collective" environmental impact.

Individual activities did not cause global warming, and individual responses will not really solve the problem. Responses to the challenges of the Anthropocene must be comprehensive and must be collective. When we begin working together with nature and one another, then we can begin to take control of our own lives. Small scale collective solutions are the only logical response. Collective actions that could produce meaningful results to reduce the impact of global warming might include: promoting public transportation or neighborhood ride sharing; building and maintaining local laundromats instead of every household having their own appliances; requiring that products be built for longevity, limiting car purchases to once every ten years to break our addictions to always having "new" this or that; or free scooters financed by local political entities. There is no shortage of creative ideas – only the desire and motivation to realize them.

Going forward, defenders of the planet must continue to expose the continuing destructive practices of capitalism at the same time that new alternative earth-compatible systems are established. Capitalists will do everything they can to maintain the current system that benefits them so richly. As Uri Gordon warned, holding their feet to the fire is essential: "The anticipation of establishment responses to collapse is crucial if anarchists and their allies are to remain ahead of the game, rather than merely reactive, considering that hierarchical institutions are already reconditioning themselves to govern collapse,"[38] Nothing will ever be the same again, and it is unlikely that maintaining current practices will never be more than window dressing covering up the ugliness of environmental destruction. More radical alternatives to capitalism must be explored, designed, and implemented.

Reciprocity, solidarity, and collective action

At the heart of anarchist theory is a belief in the power of people to organize their political lives collectively and democratically. New thinking, new skills, and new political and economic structures will be needed to confront existing power structures, and at the same time, develop meaningful problem solving alternatives outside of existing structures.

> Anarchist organization doesn't seek to replace top-down state mechanisms by standing in for them; rather, it replaces them with people building what they need for themselves, free from coercion or the imposition of authority. Rather than proceeding from a centralized polity, social organization is conceived through local voluntary groupings that maintain autonomy as a decentralized system of self-governed communes of all sizes and degrees that coordinates activities and networks for all possible purposes through free federation.[39]

Most anarchist theorists support grass-roots, local decision-making practices, often through consensus deliberation. The commune, the collective, the soviet, the kibbutz, the affinity group, the cooperative – all are models to achieve the same end – face-to-face decentralized decision-making.

There are many challenges of a political, economic, environmental, and interpersonal nature to be considered in learning how to work with others. It will be especially difficult to organize collective action in a climate of fear and uncertainty. How we proceed will likely depend upon the environment in which various collectives will function and what scale of collective action is feasible. The good news is that we can begin right now to develop alternative institutions and practice the social and political skills of cooperation. These skills will be sorely needed.

While a certain segment of the population is promoting armed survivalist camps deep in rural areas, gated communities and retrofitted underground missile silos in South Dakota for the wealthy, hoarding of foodstuffs and bullets, and similar isolationist schemes, those plans will not likely be sustainable over time. They are only temporary arks designed for maintaining a failing lifestyle and retreating from reality. Isolated communities will not be able to survive long, however, without establishing communication and cooperation with other communities. We will need each other, according to Kathryn Yusoff and Jennifer Gabrys: "Resilience and preparedness ... may not be best realized – or imagined – through depoliticized capsules for survival, but rather through more thorough-going encounters with the social and political connections that make survival and adaptation possible – and ethical."[40] Rather than hiding fearfully, what is needed is the creation of new living patterns.

Across the millennia, survival in communities has always been central to the human enterprise. We cannot survive alone. A people who pay no attention to the context and consequences of their actions, to their alienation from others and the planet will not be able to fulfill themselves, nor change the world for the better. A people who do not nurture context – whether it be a human context or an environmental/ecological context or the context of work or of social organization – those people will not survive. We must ask ourselves each day, everywhere we go, how am I connected? What is my context? How am I meshed with others and with the natural world?

The most important network for survival is, of course, the earth itself and the integrity of its diverse ecosystems and bioregions. To make real our connections to the natural world will require a change in orientation to learn to appreciate the intricacies of our ecological linkages. Timothy Morton chose the term "solidarity" to describe this connection between the human and the natural world: "*Ecological* awareness is knowing that there are a bewildering variety of scales, temporal and spatial, and that the human ones are only a very narrow region of a much larger and necessarily inconsistent and varied scalar possibility space, and that the human scale is not the top scale."[41] Solidarity is not just for people anymore – there are other beings, nonhumans, in our world. What does solidarity with nature mean? In a discussion of Kropotkin's theory of mutual aid and solidarity, Morton turned to the idea of "neighbor." "Nonhumans are being thought of as neighbors, a concept far more intense than thinking them as 'companion species,' or as being under our stewardship."[42]

> An environment is not a neutral, empty box, but an ocean filled with currents and surges.
>
> It's not just that you can have solidarity with nonhumans. It's that solidarity implies nonhumans. Solidarity *requires* nonhumans.
>
> Solidarity just is solidarity with nonhumans.[43]

Nature will be a major partner in every collective enterprise, an equal partner in all decisions, all activities, and all actions – as it should be.

I'd like to share a couple of examples of environmental challenges that will require complex decision-making in the interface between the natural world and the human residents in my own state of Minnesota. The conflicts here – and all across the country – are the same: human interests versus the environment. The northern part of the state is covered with vast forests of pine and birch. The region is colloquially called the Iron Range, and the only major economic activity in the North Woods for over a hundred years was iron ore mining and logging. Ever since the collapse of the iron mining industry toward the end of the twentieth century (because the ore ran out), the region experienced a depressed economy. To revitalize the area, unions have been advocating for decades for new jobs with excellent wages. They are specifically demanding jobs in mineral extraction industries because the wages are so high. The jobs the unions are promoting, however, will damage the environment in perpetuity.

Recently new voices have entered the political dialogue – environmentalists and wilderness preservationists advocating on behalf of the earth. International mining companies are advancing proposals for open pit copper-nickel mining – one of the most polluting industries on the planet. These plans must now take into consideration important environmental issues. Research has confirmed that this industry could potentially damage sensitive local ecosystems for centuries to come. What if the runoff from the mining pollutes the waters of Lake Superior and/or the pristine lakes and forests of the Boundary Waters Canoe Area Wilderness – one of the most popular wilderness areas in the entire country? Both are vast sources of clean water. In addition, economists have determined that industries supporting recreation in the region will create more jobs than the mining projects, but those jobs are not as well-paying. Like nature itself, decision-making will become more complex. What will win out – jobs, high-paying jobs, clean water, or minerals for electronics? What would nature say?

In another Minnesota community, a manufacturing plant was recently shut down because the children of workers had been poisoned by lead carried into their homes by their parents. The company owning the plant had supposedly been working on determining how the lead got out of the plant and into the bodies of children. After extended negotiations and no resolution of the problem the state shut the plant down. The three hundred workers want the plant re-opened because of their jobs; the company wants the plant opened so they don't go broke. At stake is the health of children. The issues uncovered here will be repeated worldwide as people become more sensitive to the pollution affecting them. Battle lines were drawn between child safety, a healthy environment, jobs, and profits without social responsibility. Who, then, will decide? Local communities, corporations, unions, or should the state or federal government weigh in? What would you do?

There are psychological hurdles to be overcome for people to engage meaningfully in collective work. Psychologist Elise Amel and others explored the psychological barriers and resistances that would have to be overcome if people are to work effectively together on climate activism. "Human beings are reticent to change their behavior even under the most compelling of circumstances, and environmental dangers do not tend to arouse the kind of urgency that motivates individuals to act. Mass transformation of unsustainable systems will be even more difficult than shifting individual behaviors, for unlike ants and bees, human are not well equipped to coordinate behavior for common benefit."[44] Further, humans are ill-equipped to detect largely invisible and gradually worsening ecological problems such as climate change or species extinction.[45]

A further problem arises when collective needs compete with individual desires, but the authors suggest there are three important ways collectives can coordinate community efforts to address conflicting motivations: "Working together to conserve a common-pool resource is difficult in the absence of enforceable limits on who can access the resource, strong social connections among community members, and opportunities for face-to-face communication."[46] All are essential principles of anarchist cooperative organizations – mutually agreed upon rules, sociality, and a dialogic environment.

There are human psychological needs that anarchist collectives will need to address such as safety, security, a desire for stability (this will be difficult in an environmentally unstable reality), and how to address the deep fears of impending death. Strong group affiliation will go a long way toward allaying some of these concerns. A further challenge will be how to help people be comfortable with change, knowing that humans don't like change. Success will depend on flexibility and adaptability and openness to change. The authors recommended several psychological tools to enhance the likelihood individual change will be welcomed and embraced: "Emphasize current and local impacts, creating incentives that increase the short-term rewards of a sustainable action, and encouraging social modeling to reset the perceived social norm around a pro-environmental behavior."[47] Encouraging people to engage in collective actions and dialogues is critical. What enhances the likelihood of collective engagement is "alignment with social identity." [48]

Decentralization
thinking and acting locally

Decentralization of power structures and decision-making are key concepts in anarchist political theory. Responses to the Anthropocene will be best undertaken at a local level. By working locally with people we know (or will soon get to know), we also personalize global warming, take responsibility for it, make it ours. This face-to-face interaction is critical in building intimate communities, according to cultural theorist Stevphen Shukaitis:

> In order for political speech to cause affective resonance, conditions need to exist for the constituted audience to be able to identify with those who are expressing them, to possess a capacity to affect and be affected. ... The continual generation of new publics, of new forms of the resonance of ideas and relations, is the process of affective composition, whether through the forming of publics through theater or any other of the possible means.[49]

Centralized power ignores the local, dictating from a distance without adequate information about local context or participation by those affected. It is an abusive model. By keeping our political power close to us, we can begin to revolutionize our responses; and we can take action in a more timely and effective manner. Anarchist theorist Brian Tokar pointed out the political power inherent in local action.

> At their best, local solutions to social and environmental problems may be more amenable to an open and accessible democratic process, and their implementation can remain more accountable to those most affected by the outcomes. Local measures can help build closer relationships among neighbors

and strengthen the capacity for self-reliance in a time of increasingly extreme climate-related disruptions.⁵⁰

At the local level hope and confidence can begin. Ecologies and the problems they present for solution are all local; we live in a wet, a cold, or a dry place; or on the seashore, a farm, or on a mountain top; or maybe in a highrise in a city center. In all our ecological neighborhoods there are actions we can take working together as communities: neighborhoods, block clubs, city or county-wide, eventually regional networks. Environmental engineer Patrick Hayden made the case for prioritizing the local: "while existing ecological problems undoubtedly present a danger to the entire planet, a micropolitical focus on the particular needs and interests of diverse local habitats and inhabitants in light of the available knowledge of ecological conditions will perhaps better contribute to the creation of effective ecopolitical interventions than will a focus solely from a unitary, large-scale framework."⁵¹

Economic decentralization is also critical to reducing the pressures on the environment. In an article exploring the linkages between energy creation and consumption, the global economy, and the role of technology, professor of human ecology Alf Hornborg concluded that if we truly wish to create a new world, we need to design a new money economy prioritizing low technology and local production and consumption. By decentralizing the economy, capitalism can be successfully undermined and replaced with an economy based on local environmental limits on energy consumption.

> The general consensus seems to be that the problem of climate change is just a question of replacing one energy technology with another. But a historical view reveals that the very idea of technology is inextricably intertwined with capital accumulation. And as such, it is not as easy to redesign as we like to think. Shifting the main energy technology is not just a matter of replacing infrastructure – it means transforming the economic world order. ... Introducing special money that can only be used to buy goods produced locally would be a genuine spanner in the wheel of globalisation.
>
> Solar power will no doubt be a vital component of humanity's future, but not as long as we allow the logic of the world market to make it profitable to transport essential goods halfway around the world. The current blind faith in technology will not save us. For the planet to stand any chance, the global economy must be redesigned.⁵²

Direct action gets the goods

Call it do-it-yourself politics. Call it taking matters into your own hands. Direct action is central to anarchist theory and practice. American anarchist Voltairine de Cleyre made the case that direct action is politically empowering by breaking decision-making dependency on majoritarian decision-making processes, and replacing them with collective processes of deliberation.

> But the evil of pinning faith to indirect action is far greater than any such minor results. The main evil is that it destroys initiative, quenches the individual rebellious spirit, teaches people to rely on someone else to do for them what they should do for themselves, what they alone can do for themselves; finally renders organic the anomalous idea that by massing supineness together until a majority is acquired, then, through the peculiar magic of that majority, this supineness is to be transformed into energy.[53]

Murray Bookchin described anarchist direct action in this way: for anarchists, direct action *"is not a 'tactic'... it is a moral principle, an ideal, a sensibility. It should imbue every aspect of our lives and behaviour and outlook."*[54] We can no longer afford to be just observers, we must become active participants to play a part in the solutions. Unfortunately, predictions of catastrophic consequences linked to global warming reduce human beings to a fearful state, resulting in paralysis. This is when taking action can break the paralysis and allow us to move on to problem-solving and empowerment. The challenge is to transform that fear into action. Hope gets actualized in action. "Anarchists reject the view that society is static and that people's consciousness, values, ideas and ideals cannot be changed. Far from it and anarchists support direct action *because* it actively encourages the transformation of those who use it. Direct action is the means of creating a new consciousness, a means of self-liberation from the chains placed around our minds, emotions and spirits by hierarchy and oppression.[55]

Direct action also serves to build and reinforce collective empowerment and build internal solidarity to achieve desired goals, according to Rob Sparrow. "Where it succeeds, direct action shows that people can control their own lives. ... We can see here that direct action and anarchist organisation are in fact two sides of the same coin. When we demonstrate the success of one we demonstrate the reality of the other."[56] Sparrow pointed out the tactical creativity and flexibility of programs that can result from localized direct action activities. They can take the form of organized protests against abuses of power, as well as build alternative cultural institutions.

> Examples of direct action include blockades, pickets, sabotage, squatting, tree spiking, lockouts, occupations, rolling strikes, slow downs, the revolutionary general strike. In the community it involves, amongst other things, establishing our own organisations such as food co-ops and community access radio and tv to provide for our social needs, blocking the freeway developments which divide and poison our communities and taking and squatting the houses that we need to live in. In the forests, direct action interposes our bodies, our will and our ingenuity between wilderness and those who would destroy it and acts against the profits of the organisations which direct the exploitation of nature and against those organisations themselves. In industry and in the workplace direct action aims either to extend workers control or to directly attack the profits of the employers.[57]

Taking action can also mitigate fear.

Anarchist practice for international collaboration

So far, I have emphasized the importance of local, decentralized political action. But the hyperobject of global warming also mandates a need for international communication and collaboration. Global warming is a planet-wide crisis and we cannot isolate ourselves from other communities and the global challenges facing all of us collectively. Forest fires in Canada affected air quality in the United States; the fallout from Chernobyl polluted vast areas of Belarus; farm runoff in Minnesota created huge dead zones in the Gulf of Mexico. We are all connected, and international agreements and actions will be necessary, if only to monitor progress toward mitigation – assuming, of course, that there is an international commitment to taking action. The highly-touted and inspiring Paris Agreement raised hopes that we could begin to move quickly ahead to address the problems confronting us. The international buy-in was nothing short of astonishing. But the truth is, there are currently no teeth in the agreement and we are still waiting for a global commitment to become reality.

There will be a need for ongoing deliberations and resolutions, as well as a mechanism for future monitoring and encouraging commitments. David Wallace-Wells, in *The Uninhabitable Earth* talked about "cascades" of change that are coming to describe the complexity and interactivity propelling the progress of global warming. Global attentiveness to these changes is an imperative. The concern is that we could fall into despair, believing that the problems are too big to tackle. This is where international cooperative efforts could buoy the human spirit, and build consensus to generate forward movement. A remarkable global collaboration recently came to light, when it was discovered that despite an international treaty to stop producing CFSs that created the hole in the ozone layer, a factory somewhere was continuing production. Thanks to a global monitoring system, scientists identified a single factory in China as the violator of the international treaty.

The concept of a federated cooperative structure has been a staple in anarchist political thought, and there are many excellent examples of historical as well as contemporary national and international coordinated activities that respect the democratic engagement of all members. These include the United Nations, the International of Anarchist Federations, federal and state credit unions in the United States, and the Mondragon Corporation, a federation of worker cooperatives in Spain founded in the 1950s. The National Cooperative Business Association (founded in 1916 as the Cooperative League of the United States of America), is a United States membership organization for cooperatives, businesses that are jointly owned and democratically controlled. There is no shortage of models for generating large scale, international cooperative action.

While we can't see global warming, we can get our thinking and our actions attuned to understanding the phenomenon. While more obvious effects of global warming may not manifest for decades or even a century or more, that change will come is certain and action now is crucial to mitigate the worst of the outcomes. Wallace-Wells writes: "There are many features of climate change – its size, its scope, its brutality … that, alone, satisfy this definition [hyperobject]; together they might elevate it into a higher and more incomprehensible conceptual category yet. But time is perhaps the most mind-bending feature, the worst outcomes arriving so long from now that we reflexively discount their reality."[58]

According to philosopher Peter Sloterdijk, to meaningfully address global warming also requires human solidarity on a planetary scale. To achieve solidarity with the earth will require solidarity with other human beings, and achieving solidarity with other humans will require overcoming the ways that religion, race, international trade, and our human tendency toward tribalism all serve to divide us rather than bind us together.[59]

How do we reach this new awareness and appreciation of the other? Celebrating our diversity is a necessary precursor to overcoming tribalism and uniting people and the environment. It should be obvious, but it bears emphasizing that diversity is the primary axiom of all of nature. As we mourn the loss of one species after another in the Sixth Extinction, the importance of diversity for our very survival is growing more apparent. All monocultures are destructive of life. Monocultures of corn kill the soil and attract and breed hostile predators, leaving the whole species vulnerable to annihilation. Monocultures of animals destroy habitat, their populations crashing as the depleted environments are no longer able to sustain them. Monocultures of economic production create corporate monopoly and massive exploitation of limited resources. A healthy community is diverse – a diversity of life forms, of ideas, of economic and creative energies. We must address the social conundrum of intolerance and the other ways in which we destroy instead of appreciate the other if we are to save the planet. Survival of one is survival of all. The contemporary awareness of global warming, has, in its way, provided an excellent opportunity to bring us together, to challenge our xenophobia and our pre-occupation with our own, narrow worlds. Can we achieve solidarity with all the beings on the planet, and begin to see ourselves as species acting together? Sloterdijk believed that the environment movement can counter tribalism and bring us together for collective survival.

> The naïve supposition of a potential openness of all to all is taken *ad absurdum* by the facts of globalization. ... What characterized "all people" without exception, "by nature" until very recently was their common and universal inclination to blamelessly ignore the vast majority of people outside of their own ethnic container. ... One of the outstanding mental effects of globalization, however, is the fact that it has elevated the greatest anthropological improbability – constantly taking into account the distant other, the stranger to one's container – to the new norm.[60]

Thinking internationally will be critical to coordinate responses, but we have major hurdles to overcome: differences in economic development; the values articulated about nature in various cultures, including religious cultures; the push and pull between nation states over control of resources; and even variations in environmental damages. Achieving this new sense of human globalization will require major shifts in thinking. Sloterdijk optimistically predicted that this transition will ultimately take place: "[T]he progressive consumption of the biosphere along with the pollution of water, air and soil changes 'humanity' willy nilly into an ecological interest group whose contemplation and dialogue must bring forth a new, far-sighted culture of reason."[61]

Globalization, unfortunately, has a dark underbelly that might work against the solidarity Sloterdijk proposed. The so-called "age of discovery" ushered in imperialism as a global phenomenon. European, and later, United States, aggressions worldwide were followed by the exploitation and brutal appropriation of natural resources that left much of the world and its resources subsumed under the shadow of the most powerful nations.

Alternative collaborative international models must replace these imperialist models. Unfortunately, our fears might lead us to trust the centralized corporate and political power structures that have generated the environmental problems in the first place. Again, decentralization is a powerful antidote to centralized command and control power structures. Michael Truscello urged caution and warned against new global forms that would only replicate existing "arborescent" (tree-like) power. "Any global response to environmental crises is more likely to produce arborescent power structures than it is to produce open multiplicities. ... The prolonged emancipation from this rule of arborescent thought will require an 'unprecedented proliferation of "open systems" attuned to "diverse local habitats and inhabitants," not a one world order of resistance."[62]

Arborescent is a term coined by French philosophers Gilles Deleuze and Félix Guattari, describing formal rigid, hierarchical structures of power that preclude freedom and flexibility, and stifles creativity. Truscello called instead for a "post-anarchist ecology" that assumes a horizontal, decentralized model for anarchist organization – a vast network of interconnected decentralized nodes holding the web of life in its balance.

An international model must also commit to share and develop resources for the good of all. While some have continued to nurture dreams of new energy source or bigger and better technologies, an interim plan must be put in place to redistribute existing resources worldwide and set limits on the exploitation of those resources. In order to do this, Mulford Sibley argued, we need to accept "such principles as public ownership of land and natural resources, regulation of migration on a world scale, establishment of world-administered food reserves, and fundamental changes of consumer habits within the developed nations."[63]

Another World Is Possible and Necessary

Resisting capitalism and actively building a collaborative future is crucial for anarchist practice, but it is not quite enough. Another political and economic system must replace it – a system in balance with the needs and demands of nature. This new world will incorporate anarchist principles outlined previously. This world will be created anew – not in one wave of a magic wand, but as part of an ongoing project. It will not be easy, and uncertainty will be the order of every day. The creation of the new world will be open to revisions and redirections and reconsiderations and even reversals. "[A]narchism is not about drafting sociopolitical blueprints for the future, nor does it trace a line or provide a model. Prefiguration should not be confused as predetermination, as anarchists are more concerned with identifying social tendencies, where the focus is on the possibilities that can be realized in the here and now."[64]

Regeneration cannot be undertaken until there is the recognition that we have an ethical responsibility to repair the damage done. Daniel Wahl outlined the challenge: "Creating regenerative systems is not simply a technical, economic, ecological or social shift: it has to go hand-in-hand with an underlying shift in the way we think about ourselves, our relationships with each other and with life as a whole."[65] This will require an ethical reorientation in our thinking as outlined in Emmanuel Levinas' ethical formulations in the chapter on the sublime. According to Irish philosopher Kieran Cashell:

> Levinas developed a principle of irreducible alterity, with its key claim that the *other person* remains irreducibly prior to, and transcendent of, any conception of the self. ... Anarchism is redefined in this context as an ethically-motivated political practice of 'infinite responsibility' – characterized by an imperative (or 'call') issued from the Other that challenges and disturbs the epistemic and moral authority (as well as the security) of foundational egocentric subject categories.[66]

Creating a new world requires an openness to change and reinforcement of the belief in the power of people to embrace change. The structural flexibility of an anarchist culture makes change easier because there is face-to-face communication and collaborative efforts in problem-solving. As Kenneth Heilman and Russel Donda reminded us, all people have a capacity for change: "It is an individual's ability to diverge from what is familiar and move beyond the known into a new understanding which is the essence of creativity, and that which gives rise to advancement."[67] People who believe they can change are more likely to act politically on that belief. Social anarchist art can actualize this ability to change our worldview by building a bridge to an alternative vision that incorporates an ongoing process.

Only change that begins in a real world context and moves the viewer to a realizable future will empower people to act on their hopes and dreams. To change, people must imagine new ways of doing and being and believe they can make the transition. Portraying the process of realizing goals by portraying "how" to change will create a new world, will make the future achievable, not merely utopian fantasy. The future must be linked with the present, but portray a different path forward – one filled with passion for developing a relationship to meet the needs of all beings.

Nourishing the power of the imagination is an extremely important tool for inspiring political change. Here is where artists will play a key political role. This was the essence of the surrealist idea. The free play of imagination is a necessary prerequisite to opening up belief in possibilities and is critical to the creation of political art. Artist Elizam Escobar argued for prioritizing a more political imagination:

> If art is to become a force for social change, it must take its strength from the *politics of art,* art's own way of affecting both the world and the political directly. The politics of art will happen only if the power of the imagination is able to create a symbolic relationship between those who participate, the artwork and the concrete world, always understanding the work of art's sovereignty (or relative autonomy) in relation to concrete reality.[68]

Political philosopher Richard Gilman-Opalsky reflected on the relationship between political power and imaginative power:

> What people want is *real* power, not imaginary power. But I think that imaginary power is real power. And this where I think that art as something more than epiphany is very important, because if you can't imagine other ways of life, other forms of life, then you cannot demand them. You cannot build what you cannot imagine. ... In this way, art is part of the production of power. Its imaginary power can become real power, or already is real power, in the sense that it helps us to think about *real* possibilities beyond the existing realities.[69]

As Theodor Adorno observed, it is the dialectic between the real in which we are grounded and the possible to which we aspire that creates the tension characterizing great art and politics. The biggest challenge is bringing the real and the possible together. The possible, of course, is always latent in the real. The challenge to art is to awaken that potential, to visualize, and to vitalize it. The new world must be a realizable world, not abstract or idealized.

Nimble, flexible, creative, the freedom of visioning – these are skills that will be needed in the future to respond effectively to the many problems and challenges we will face as existing systems and structures fail, and system responses are either too slow or ineffective. These visionary skills can best be nourished in an environment of cooperation, openness to seeking alternatives, and commitment.

REWORLDING
LISTENING TO THE VOICES

> Hope has two beautiful daughters; their names are
> Anger and Courage. Anger at the way things are, and
> Courage to see that they do not remain as they are.
> – Augustine of Hippo

CONCLUSIONS ARE ABOUT SUMMING UP THE INFORMATION PRESENTED throughout a book, pointing out all the major highlights, and tying everything up in a nice package to take home and put on the shelf. This book, unfortunately, has no neat and tidy answers, no clever or brilliant conclusions. The challenges of the Anthropocene are tasks to be taken up by all of us, in any way we can, to save ourselves, future generations of humankind and the other beings with whom we currently share the world. So this conclusion is really only a beginning.

This has been a dark book, and any conclusions must take into account the power of dark feelings. One might ask at this point whether stirring up fear over the realities of the Anthropocene throughout this book was a wise idea. I could have written a different book – one filled with cheerful platitudes that "something" will save us. But that was not to be, because the stakes are so high and the reality so stark that I felt awakening our sense of self-survival was critical. In defense of looking on the dark side of the challenges facing us as species, I share with you the words of an environmental writer who apologized (pseudononymously) for his failure to adequately warn us of the impending catastrophe.

> In all of these jobs, my bosses told me I shouldn't scare the horses; I shouldn't tell the full truth of the encroaching horrors I researched all day. Otherwise, people might give up or – in other words – stop sending us money.
> I complied. I wanted the money. And I'm sorry. ...
> When did reality become so unpalatable that we shouldn't tell people about it? Since when did so many of us need to be treated like children? Why do those of us who know about these things have to disassemble, distract or blatantly lie about the state of our world?[1]

We must ultimately come to grips with the damage we have done in creating this Anthropocene. We can't avoid the pain and suffering we have created. The way forward is through our grief – a grief that demands we own up to our actual and potential destructiveness and the disruptions to plant and animal life we have wrought. Are we ready to recognize what we have done and what we still continue to do? That the capitalist economic system with its myth of fulfilling individual desires and the continued unlimited growth we uphold is destroying the earth? That we all are participants in the potential death of the planet? Are we strong enough to confront the terror we have created? Do we have the heart to take up the fight, not against the others, but with the others to live?

Perhaps another conclusion we can come to is that it is better to know and move forward with our eyes wide open, with our hearts filled with sorrow and remorse, and the resolve to take a different way forward.

Lessons from the Great Flood

I want to return one last time to the centrality of water – the foundation of all life – and to the long history of water narratives of destruction, of compassion, and hope embedded in water as sustainer and source of life. Tales of a great flood are ubiquitous in cultures around the globe and across time. In one telling, the Great Flood was sent by the god as punishment for the sins of humankind. In the tale, Noah is charged with building an ark and bringing representatives of all the beasts and beings of the planet into the ark to save them. We do not know what Noah thought of the creatures he took onboard, but he built the ship strongly and prepared carefully for a journey into an unknown, uncertain future, gathering together all of the beings to replenish

Fig. 25 Clifford Harper, *The Flood*

the earth after the waters subsided. Compassion must have played a role in his meticulous accounting as earth's beings climbed aboard to their future. A similar tale of the building of an ark appeared in Hindu mythology. Matsya, a fish avatar of the god Vishnu, warns of a flood, an ark is built, and is filled with provisions for future life. Some beings, of course, were left behind, and the stories don't tell us about how humanity and those other beings felt about the catastrophic loss of life.

It seems that the gods of destruction and chaos may have also been compassionate gods. They assigned the task of saving the earth to good people, granting humankind another chance at living differently in the world.

There is another empowering addendum to this ancient tale from which we can take heart. It is a tale of active hope emerging from disaster. The inundation returned the earth to its original state of chaos, out of which a new, and presumably better world was again created. The hope embedded in the story of the Great Flood is that there are survivors – all of the beings together surviving to start anew.

We are now living through the Great Flood of the climate crisis. The flood is upon us along with the potential for a new chaos. Out of the chaos we have an opportunity to create anew, to reworld this magnificent earth. This time, the ark is our entire planet and all living beings.

Reworlding

I chose the word "reworlding" because it is a new word – when I first looked up its definition, it didn't appear in any dictionaries. The word worlding, which came the closest, had a very narrow philosophical application that had nothing to do with the idea of our world, our earth, our planet. Reworlding in the sense I use it is about making a new world, or at least about linking our actions with the world. The term presupposes active engagement, not distancing. Later searches turned up references to the word reworlding in a new platform or a new interactive game on the internets that hinted at creating new intergalactic worlds. While futuristic, off-planet fantasies can be creative and offer potential for exciting world building, they tend to be utopian and unrealistic. I'm more interested in putting those creative energies to work in this world we call earth and reworlding a cooperative, interactive future here. It is new word for a new time. To "world" again, to begin again, has the potential for us to live here in a new way.

Reworlding is also an active verb, and like all words of doing and making, it is an ongoing process – a conscious political and ethical effort requiring attention and engagement in our human communities and with the animate and inanimate beings around us. Reworlding is part resistance, part rejection of the ways of contemporary life that are killing all life, and part resilience – learning new ways to live strongly in a life-affirming manner. Most of all, it is grounded in material actions. Reworlding requires that we respond with engagement, with actually doing something differently. It means doing our living with this planet in a new way. We cannot just think of ourselves as manipulating again the physical world in a new way, but changing our thinking about how we co-evolve.

That little prefix "re" interested me, too. It occurred originally in loanwords from Latin and plays an important place in the English language. It can mean "again" or "again and again" to indicate repetition, or it can indicate a re-do, a starting over as in regenerate; restore; repair; or even revert, as in going back to an original state. Being able to begin again allows for a renewed hope that we can move forward. Throughout the book I've explored such revitalizing concepts as rewilding, recognizing, reenchantment, and especially reimagining the narratives of our collective lives.

Re-wilding

Seeking to recapture what we have lost in the natural world is an exercise in learning again the foundations of life. There is no wild as such anymore, because nothing escapes our touch. Our water and air pollution, for example, is manifested in acidification in lakes, rivers, and oceans. Our native forests suffer drought and are regularly swept by invasive plant and animal diseases. Rewilding is a commitment to return the planet as best as we can to its previous health – to repair, to make reparations. Even though we know we can never return to an idealized Garden of Eden, we can make our garden grow and thrive once again. Gardening, as Indigenous environmental activist Winona LaDuke reminded us in a recent Facebook comment, "is like a courtship."[2]

There is an active interaction required to love the earth back to life. We know we can do this, because we've fixed some of our mistakes in the past.

Re-cognition

Our very thinking – the use of our cognitive abilities – is changing as we awaken from the nightmare of exploitation of the natural world and learn to see our world in new, richer ways. Recognition means to re-know, re-think, re-conceptualize. It is also the process of recapturing a lost, forgotten, or repressed awareness. It is a matter of reconnecting with the earth and with one another. Recognizing the complexity of the living planet is key. We know so little about how each being affects the whole (including how it affects us, who are also part of the whole). Nevertheless, we have an obligation to learn from these others and to maintain the balance and health of the whole. We can learn from past errors and use the knowledge of those unintended consequences to act proactively in the future. Science can help us with this task. We can begin by looking around our special place – our region, our community, our ecosystem.

Re-sponsibility

Nourishing the ability to respond, according to Wallace-Wells, lies at the heart of our reworlding: "[T]here is no mysticism required to interpret or command the fate of the earth. Only an acceptance of responsibility."[3] Responsibility is central to a new way of thinking, giving up our sense of control and superiority and our dominance of nature. Instead, we can move closer to and embrace our connections with the natural world. To be responsible is to simply recognize the reality and equality of the others living in our world, and begin a dialogue.

Re-enchantment

Re-enchantment is a matter of visioning, of learning to see in a new way. Given the uncertainty and gravity of the future facing humankind, the need to inspire and engage all of us in this process of reworlding is crucial. Reworlding means to overcome our alienation, honor our deep linkages with the natural world, and return to our true "home" with other beings. Reworlding is the art of being home in a world re-illuminated with its power and its magic.

Here is where artists will play a central role – revising and re-imagining our visions, shaping our future, opening up new narratives celebrating change, and making those new visions go viral. In these dark times, art and artists have powerful messages for us – messages gleaned from the past and reimagined for the future. So what does art of the Anthropocene look like in this time of transition? Some artists are reflecting the fear and terror of the climate crisis and imagining a new world. Increasing numbers of artists are addressing the impending catastrophe head on. Some are visualizing new relationships. Some artists are calling for empathy, in an effort to heighten awareness of the losses created in the Sixth Extinction. Others try to alert us by appealing to our human drive for self-preservation, sending us warnings of impending threats. Other artists call us to action, moving us through and beyond fear – creating propaganda for the planet. Still others will stir our deep human passion and the desire to connect with the life force around us.

Listening to the Voices

In *Being Ecological,* Timothy Morton proposed a way we can move forward by exploring new pathways, new ways of knowing and being with nature. We can begin by paying close attention, by opening our senses to the world around us, by listening: "All beings have voices, they also have agency. Are we listening? What are other beings saying?"[4] Listening? But how can we listen? Of course, the trees don't speak. It is here that our scientific knowledge, our creativity, and our desire to live can begin to shape our vision of the health of the whole. Take trees, for example. Trees won't actually "talk" to us, but we can learn what a tree does, its role in the complex system of life, and how it affects us. Knowing what the tree does, how it is acting out its desire to live, can be a kind of listening. Knowing nature better and more deeply will awaken us to the vitality of the world. Trees are the lungs that give us all breath. They breathe for us all through their relationships with oxygen and carbon dioxide. Because of this knowing we are listening. We cannot ignore the voices asking to live, because they are asking on our behalf as well. The tree's voice and our voice resonate with the intelligence of the whole. In the poem "Lost", David Wagoner reminded us:

> If what a tree or a bush does is lost on you,
> You are surely lost. Stand still. The forest knows
> Where you are. You must let it find you.[5]

All the others in our environment – the trees, the rivers and lakes, the bears and wolves, the bees and the ants – are acting and interacting in our shared ecosystems every day. Like the ancient peoples who imagined the world as alive and activated by these beings, we too can experience the world as alive again through our renewed knowledge. What images, icons, symbols, and stories will we create? The ancient ones imagined active gods and goddesses of creation, the seas, the weather. They knew their world intimately – they recognized and honored the earth as an active force in their lives, experiencing and communicating nature's complexity by assigning personhood and agency to all elements in their environment. Every culture recognized the life givers of nature itself, the creators, like the Greek Titans who brought the world out of the chaos. Principles of fertility and abundance were recognized in beings like the Hopi Kokopeli, the Inca Pachamama, the Greek Demeter and Persephone who ruled over the seasons and agricultural plenty, the Inuit Nanuk, a bear god who represented food, and the Inuit goddess Sedna who filled oceans with the beings of the waters. The Hindu pantheon is a rich community of earth deities. There were ocean gods, river gods, and swamp beings, along with a host of animal beings peopling their world – eagles, bears, ravens, and coyotes. There were gods and goddesses of the weather, of local places, of forests, and land formations like mountains and volcanoes. Hawaiians still have great respect for Pele, the goddess of fire and the vagaries of weather.

We have forgotten all of this knowledge in our mad rush to dominate and exploit the other beings. Now we need to re-member, to gather together the pieces of the whole that ensure our own survival. We are, in the end, equal partners in maintaining our planet earth.

Listening to the voices is, at its heart, a matter of respecting and honoring boundaries. A study by the Stockholm Resilience Centre (2009) summarized by Daniel Wahl, identified the planetary boundaries which humankind had to not only respect, but learn to live within the limits.[6] For example, the key measure controlling climate change is the amount of carbon dioxide in the atmosphere – a limit we cannot exceed. We are learning about other limits: the necessity of biodiversity from bees and insects; how the nitrogen/phosphorus cycles affects plant growth; how chemical pollution is widespread and is affecting all life; how carbon pollution is acidifying the oceans and killing creatures with shells like corals; and how fresh water sources are increasingly polluted worldwide. Disruptions in one of these systems can have cascading effects on other systems as well, and on the earth a whole. Every one of these systems and their disruptions affects humankind. As scientists better understand the human impact on these systems, our behavior will have to be adapted to ensure stability. We are still in the process of listening and learning, and even though the way forward is not yet completely clear, knowledge and deep respect for the others will forge a new world.

Hope as Action

Now that we know the truth about the Anthropocene and we have the resolve to find a way forward, there is no turning back, no way to wiggle out of our grief and our responsibilities. Overcoming our fears and our sorrows, we have to move through them to hope. Our gift to the world is our passion for living itself, the heart of hope. Our way forward is to choose to act. Action, in turn, regenerates and then magnifies our hope as we see our renewal efforts succeeding.

In *The Wake of Crows: Living and Dying in Shared Worlds,* Thom van Dooren visited several global sites of crow/human interaction. Each locale had developed its own relationship with the crows, ranging from making every effort to exterminate them, to re-introducing them into sharing environments. His study of crow/human interactions is a deep mediation on how we can communicate and care for the other beings in our world. He explored how hope and action must be embedded in the remaking of our world. Nature – the world – has much to teach us about how the passion for living is a powerful combination of resistance to dying and resilience in the face of great challenges. In his accounts of interactions with humans, the outcome for crows was often a sad and precarious one, but van Dooren continued to believe in a better future grounded in hope that we can learn new, respectful ways. Hope, as he pointed out, is not a passive matter, but one of commitment and action:

> As a mode of worlding, hope is not always grounded in this kind of deliberate action; there are many ways in which hopes shape worlds. But it is this kind of active work that interests me most strongly: hope as a work of care for the future. Much more than an anticipation of or a simple desire for a coming good, hope is an effort to care for that possibility in committed, practical and situated ways.[7]

Environmental action begins with recognizing that the other being is important – and in that recognition lies the hope for a future ongoing and respectful relationship with the other being.

> Hope is a project, a labor ... It is not a vague optimism about the future but something that must be worked on, built, shared with others so that it might take root and grow unpredictably, wildly. Hope is an ongoing effort to cultivate the conditions for a better future. ... From this perspective, hope does not preexist acts of hopefulness. Hope grows and expands as we actively care for the future. It is an iterative, cumulative, self-reinforcing work, an effort to do whatever one can – within the context of the many, always unequal constraints that shape our possibilities – to nurture a better future into being.[8]

Hope can be built on this new perspective. In *Of Wolves and Men* - a compassionate reflection on the violent and murderous human/wolf relationship across time – author Barry Lopez reminded us that as we learn a new way of seeing and listening to the others in our world, we can reimagine a relationship of care and compassion instead of death and destruction. There was a time in some indigenous cultures where the wolf was regarded as an equal being in high regard. Lopez asked us to reclaim those insights or mutual respect: "In the end, I think we are going to have to go back and look at the stories we made up when we had no reason to kill, and find some way to look the animal in the face again."[9]

But it isn't just a matter of listening – our challenge will be to transform the messages we hear into action. What does nature demand of us? What will we do, I ask? In reply, my good friend Eric Steinmetz observed, "We cannot please Nature nor should we imagine that Nature will punish us for our crimes. We mourn for the loss of these iconic animals and landscapes because they are most like us and the challenges we see ourselves achieving. Well, all right then. We will save the world we love, not for its sake, but for ours."[10]

This is the role of art – envisioning a world filled with hope and meaningful engagement with the natural world. This emerging aesthetic will encapsulate the essence of a new expanding awareness – recognizing the voices of the earth and listening to the messages from nature. Survivors of the Anthropocene will prevail by creating a new place for humanity in the re-worlding of the planet by listening, grieving, living, making art, working together, and engaging with all other survivors. Beyond grief and acknowledgment lie hope and action. From this new relationship will come new stories, new visions. Artist James Koehnline challenges us to open our hearts, put our ears to the ground, and get busy.

Fig. 26: NomadicAlternatives.org, *Forgotten Tongue*

So my question is how do we break through the "agrilogistic" programming, the old mythology, the chains of the old zodiac, the fixed set of stories we can hear and live by, so that we can engage in a more meaningful way with the rich plenitude, with all its joys and hazards, of our entanglement in the place, time, and multispecies complexities of life on Earth? Now is the time to figure this out. What kinds of stories could be the seeds of a robust co-existentialism, capable of seeing through the dazzling spell of capital and finding some solidarity with what is left of the web of life on Earth?[11]

BIBLIOGRAPHY

A Sea Ethic. http://marinebio.org/oceans/conservation/sea-ethic/

Abram, David. *The Spell of the Sensuous: Perception and Language in a More-Than-Human World.* New York: Vintage, 1996, 2017.

Adorno, Theodor W. *Aesthetic Theory.* Eds., Gretel Adorno and Rolf Tiedemann. Translated with a translator's introduction by Robert Hullot-Kentor. Minneapolis: University of Minnesota Press: 1997.

_____. *Negative Dialectics.* Trans. E. B. Ashton. London and New York: Routledge, 1973, 2004.

Aesthetics and Politics: Theodor Adorno, Walter Benjamin, Ernst Bloch, Bertolt Brecht, and George Lukacs. Afterward by Fredric Jameson. London & New York: Verso, 2007.

Ahmed, Nafeez. "Government Agency Warns Global Oil Industry Is on the Brink of a Global Meltdown." *Vice* (February 4, 2020). https://www.vice.com/en_us/article/8848g5/government-agency-warns-global-oil-industry-is-on-the-brink-of-a-meltdown

Albrecht, Glenn. "Exiting the Anthropocene and Entering the Symbiocene." *Minding Nature* (Spring 2016), 9, 2. https://www.humansandnature.org/exiting-the-anthropocene-and-entering-the-symbiocene

Albrecht, Glenn, and eight others. "Solastalgia: The Distress Caused by Environmental Change." *Australian Psychiatry* (February 2007. https://www.researchgate.net/publication/5820433_Solastalgia_The_Distress_Caused_by_Environmental_Change

Allen, Bruce. "First There were Stories: Michiko Ishimure's Narratives of Resistance and Reconciliation:" 35-57. In Estok, Simon C. and Won-Chung Kim, Eds. *East Asian Ecocriticism: A Critical Reader.* Macmillan, 2013.

Amel, Elise, Christie Manning, Britain Scott, Susan Koger. "Beyond the roots of human inaction: Fostering collective effort toward ecosystem conservation." *Science* 356 (April 21, 2017), 275-179. https://science.sciencemag.org/content/356/6335/275.abstract

Aravamudan, Srinivas. "The Catachronism of Climate Change." *Diacritics* 41.3 (2013): 6-31.

Bateson, Nora. "Warm Data to Better Meet the Complex Risks of This Era." December 7, 2018. https://norabateson.wordpress.com/2018/12/07/warm-data-to-better-meet-the-complex-risks-of-this-era

Bayles, David and Ted Orland. *Art & Fear: Observations on the Perils (and Rewards) of Artmaking.* Santa Cruz, CA and Eugene, OR: The Image Continuum, 1993).

Bendell, Jem. "Deep Adaptation: A Map for Navigating Climate Tragedy." IFLAS Occasional Paper 2. July 27, 2018. www.iflas.info

Berg, Ina. "Towards a conceptualisation of the sea: artefacts, iconography and meaning." In Giorgas Vavouranakis, Ed., *The Seascape in Aegean History.* Monographs of the Danish Institute at Athens, Volume 14: 119-137.

Berger, Edmund. "The Anthropocene and the End of Postmodernism." *Synthetic Zero.* April 2, 2017. https://syntheticzero.net/2017/04/02/the-anthropocene-and-the-end-of-postmodernism/

Berkman, Alexander, Ed. *Selected Works of Voltairine de Cleyre.* New York: Mother Earth Publishing, 1914.

Berleant, Arnold. *Aesthetics and Environment: Theme and Variations on Art and Culture.* Ashgate Publishing: England and Vermont, 2005.

Berman, Morris. *The Reenchantment of the World.* Ithaca, New York: Cornell University Press, 1981.

Berns, Jasper. "Between the Devil and the Green New Deal." *Commune* 3 (Summer 2019). https://communemag.com/between-the-devil-and-the-green-new-deal

Biello, David. "Mass Extinctions Tied to Past Climate Changes" *Scientific American* (October 24, 2007). https://www.scientificamerican.com/article/mass-extinctions-tied-to-past-climate-changes

Bittle, Jake. "On the Waterfronts: Flood buyouts and thek economics of climate catastrophe," *The Baffler* 49 (January-February 2020). 125-133.

Bjornerud, Marcia. *Timefulness: How Thinking Like a Geologist Can Help Save the World.* Princeton and Oxford: Princeton University Press, 2018.

Blasdel, Alex. "A Reckoning for our Species: The Philosopher/Prophet of the Anthropocene," *The Guardian.* June 15, 2017. https://www.theguardian.com/world/2017/jun/15/timothy-morton-anthropocene-philosopher

Bloom, Brett. *Petro-subjectivity: De-Industrializing Our Sense of Self.* Aubun, IN: Breakdown Break Down Press, 2015, 2018.

Boes, Tobias and Kath Marshall. "Writing the Anthropocene: An Introduction." *Minnesota Review* 83 (2014): 60-72.

Bookchin, Murray. *The Ecology of Freedom: The Emergence and Dissolution of Hierarchy.* Oakland, CA: AK Press, 2005.

_____. *Remaking society.* Montreal: Black Rose Books, 1989.

Bradford, George. *How Deep is Deep Ecology?* Edinburgh: AK Press, 1989.

Brady, Emily. "Reassessing Aesthetic Appreciation of Nature in the Kantian Sublime." *Journal of Aesthetic Education* 46, 1 (Spring 2012): 91-109.

Brady, William J., Gantman, A. P., & Van Bavel, J. J. "Attentional Capture Helps Explain Why Moral and Emotional Content Go Viral." *Journal of Experimental Psychology: General.* Advance online publication. September 5, 2019. http://dx.doi.org/10.1037/xge0000673

Braudel, Fernand. *The Mediterranean and the Mediterranean World in the Age of Philip II.* Trans. Siaˆn Reynolds, 2 vols. (1949; London, 1972).

Broad, William J. "The Ice Was Only Part of It." *New York Times,* April 9, 2012. https://www.nytimes.com/2012/04/10/science/a-new-look-at-natures-role-in-the-titanics-sinking.html

Brown, H. Claire. "How Ohio's Chamber of Commerce Killed an Anti-Pollution Bill of Rights." *The Intercept.* August 29 2019. https://theintercept.com/2019/08/29/lake-erie-bill-of-rights-ohio/

Brown, Stephen and Anthony Patterson, Eds. *Imagining Marketing: Art, aesthetics and the avant-garde.* New York and London: Routledge, 2000.

Brunetta, Victoria and Kate O'Shea, Eds. *Durty Words: A space for dialogue, solidarity, resistance and creation.* Limerick, Ireland: Durty Books, 2019.

Bryce, Emma. "What do 10,000 Europeans know about climate change and the sea?" http://www.anthropocenemagazine.org/2017/07/what-do-10000-europeans-know-about-climate-change-and-the-sea/

Buchloh, Benjamin H.D. "Allan Sekula: Photography between Discourse and Document." In Sekula, Allan, *Fish Story.* Rotterdam: Richter Verlag, 1995: 190-200.

Buell, Lawrence. *The Environmental Imagination: Thoreau, nature writing, and the formation of American Culture.* Cambridge, MA: Harvard University Press, 1995.

Bukharin, Nicolai. *Historical Materialism: A System of Sociology*, 1921. Chapter 3. https://www.marxists.org/archive/bukharin/works/1921/histmat/index.htm

Burke, Edmund. *A Philosophical Enquiry into the Origin of Our Ideas of the Sublime and Beautiful.* Second edition, 1759. http://www.english.upenn.edu/~mgamer/Etexts/burkesublime.html

Burke, Edmund. *Philosophical Enquiry into the Sublime and Beautiful and Other Pre-Revolutionary Writings*, ed. by David Womersley, London 1998: 102.

Carlson, Allen and Arnold Berleant, Eds. *The Aesthetics of Natural Environments."* Guelf, Ontario: Broadview Press, 2004.

Carson, Rachel. *The Sea around Us*. New York and Oxford: Oxford University Press, 1989 (1950).

Cashell, Kieran. "Scared Shitless: Anarchism, Capitalism or Democracy." In Victora Brunetta and Kate O'Shea, Eds., *Durty Words: A space for dialogue, solidarity, resistance and creation*. Durty Books: Limerick Ireland, 2019.

Chakrabarty, Dipesh. "The Climate of History: Four Theses." *Critical Inquiry* 35 (2): 197-222.

Chaudhary, Ajay Singh. "It's Already Here: Left-wing climate realism and the Trump climate change memo." *N+1*. March 20, 2019. https://nplusonemag.com/online-only/online-only/its-already-here

Cheng, Xiangzhan. "Aesthetic Engagement, Ecosophy C, and Ecological Appreciation." *Contemporary Aesthetics*. 11 (2013). http://hdl.handle.net/2027/spo.7523862.0011.009

——————. "On the Four Keystones of Ecological Aesthetic Appreciation," *East Asian Ecocriticism: A Critical Reader*, eds., Simon C. Estok and Won-Chung Kim (Macmillan, 2013), 213-228.

Cho, Renee. "Climate Change Poses Challenges to Plants and Animals." Earth Institute Columbia University (February 3, 2015). https://blogs.ei.columbia.edu/2015/02/03/climate-change-poses-challenges-to-plants-and-animals/

Chung-yam PO. *Conceptualizing the Blue Frontier: The Great Qing and* the Maritime World in the Long Eighteenth Century. Reprecht-Karls-Universitat Heidelberg. http://archiv.ub.uni-heidelberg.de/volltextserver/18877/1/PhD_Dissertation_CyPO.pdf

Clark, John P. "The Dialectical Social Geography of Elisée Reclus." From *Philosophy and Geography I: Space, Place, and Environmental Ethics*. (Lanham, MD: Rowman and Littlefield, 1997), pp. 117-142. https://www.academia.edu/2540498/_The_Dialectical_Social_Geography_of_Elisee_Reclus_

Clark, John and Camille Martin, eds. *Anarchy, Geography, Modernity: Selected Writings of Elisée Reclus*. Oakland, CA: PM Press, 2013.

Clark, John. "How an Anarchist Discovered the Earth." June 27, 2018. https://www.academia.edu/36394802/_How_an_Anarchist_Discovered_the_Earth_Revised_6_27_18_

Colebrook, Claire. "Earth Felt the Wound: The Affective Divide." *Journal for Politics, Gender, and Culture*, 8(1), 45-58, 2011.

Collaborative on Health and the Environment. "Precautionary Principle: The Wingspread Statement." https://www.healthandenvironment.org/environmental-health/social-context/history/precautionary-principle-the-wingspread-statement

Cook, John, et. al. "Consensus on consensus: a synthesis of consensus estimates on human-caused global warming" *Environmental Research Letters* (2016). https://iopscience.iop.org/article/10.1088/1748-9326/11/4/048002/pdf

Critical Art Ensemble. *Aesthetic, Necropolitics, and Environmental Struggle*. New York: Autonomedia, 2018.

Crockett, Daniel. "Nature Connection Will Be the Next Big Human Trend." *Huffington Post.* October 21, 2014. http://www.huffingtonpost.co.uk/daniel-crockett/nature-connection-will-be-the-next-big-human-trend_b_5698267.html

Cunsolo, Ashlee and Neville R. Ellis. "Ecological Grief as a mental health response to climate change-related loss." *Nature Climate Change,* 8 (April 2018), 275-281.

Cutts, Steve. *Wake Up Call: End the Nightmare of Consumption.* https://www.filmsforaction.org/watch/wake-up-call-end-the-nighmare-of-consumption/

Dadejík, Ondřej and Vlastimil Zuska. "More than a Story: the Two-Dimensional Aesthetics of the Forest." *Estetika: The Central European Journal of Aesthetics,* 1, 2010: 27– 50.

Davis, Heather and Etienne Turpin, Eds. *Art in the Anthropocene: Encounters among Aesthetics, Politics, Environments and Epistemologies.* Open Humanities Press, 2015. http://openhumanitiespress.org/books/art-in-the-anthropocene

——————————————————. "Art & Death: Lives between the Fifth Assessment & the Sixth Extinction." In Heather Davis and Etienne, Turpin, Eds. *Art in the Anthropocene: Encounters among Aesthetics, Politics, Environments and Epistemologies.* Open Humanities Press, 2015. http://openhumanitiespress.org/books/art-in-the-anthropocene

Debord, Guy. *Society of the Spectacle.* Trans. and Annotated by Ken Knabb. Berkeley, CA: Bureau of Public Secrets, 2014.

Debord, Guy. *Comments on the Society of the Spectacle.* Pirate Press, 1991; Verso, 1990.

de Cleyre, Voltarine. "Direct Action." *The Anarchist Library.* https://theanarchistlibrary.org/library/voltairine-de-cleyre-direct-action

Deleuze, Gilles and Guattari, Félix. *What is Philosophy?* H. Tomlison and G. Burchill, Trans. London: Verso, 1994, 166.

Demos, T. J. "The Great Transition: The Arts and Radical System Change." *E-flux Architecture.* http://www.e-exflux.com/architecture/accumulation/122305/the-great-transition-the-arts-and-radical-system-change/

Detwiler, F., 1992. "All my relatives: persons in Oglala religion." *Religion,* 22 (3), 235–246.

"Direct Action." http://www.infoshop.org/direct-action/

Doane, Bethany. "Aesthetics, Ethics, and Objects in the Anthropocene." Review. Timothy Morton, *Hyperobjects: Philosophy and Ecology after the End of the World.* Minneapolis, MN: University of Minnesota Press, 2013.

Driessen, Henk. "A Janus-faced Sea: Contrasting Perceptions and experiences of the Mediterranean." University of Nijmegen, The Netherlands. Accessed 4/1/17. http://www.marecentre.nl/mast/documents/ArtikelHenkDriessen.pdf

Drichel, Simone. "Face to Face with the Other: Levinas versus the Postcolonial." *Levinas Studies* 7 (2012): 21-42.

Drooker, Eric. *Flood! A Novel in Pictures.* Milwaukee, OR: Dark Horse Books, 1992, 2007.

Duarte, Edivania Lopes. *Between the Uncanny and the Sublime: An Investigation of the (self-)Reflective Process in American Gothic Literature.* Master Thesis, Universiteit Utrecht, 2014.

Earth System Research Monitoring Laboratory, Global Monitoring Division. "History of atmospheric carbon dioxide from 800,000 years ago until January, 2019. https://www.esrl.noaa.gov/gmd/ccgg/trends/history.html

Eckersley, Richard. "Nihilism, fundamentalism, or activism: three responses to fears of the apocalypse." *The Futurist* [Online], January-February 2008. https://richard-eckersley.com.au/attachments/Futurist_Apocalypse_2008.pdf

Ellis, Neville and Ashlee Consulo. "Hope and Mourning in the Anthropocene: Understanding ecological grief." *The Conversation* April 4, 2018. http://theconversation.com/hope-and-mourning-in-the-anthropocene-understanding-ecological-grief-88630

Engber, Daniel. "There is No Island of Trash in the Pacific." *Slate.* September 12, 2016. http://www.slate.com/articles/health_and_science/the_next_20/2016/09/the_great_pacific_garbage_patch_was_the_myth_we_needed_to_save_our_oceans.html

Engels, Friedrich. "The Part played by Labour in the Transition from Ape to Man." Trans. Clemens Dutt. 1876. https://www.marxists.org/archive/marx/works/1876/part-played-labour/index.htm (Progress Publishers, Moscow, 1934 - English).

Escobar, Elizam. "Art of Liberation: A Vision of Freedom", in *Art on the Line, Essays by Artists about the Point Where Their Art & Activism Intersect,* ed. Jack Hirschman (Williamantic, CN: Curbstone Press, 2002), 248-249.

Estok, Simon C. and Won-Chung Kim, Eds. *East Asian Ecocriticism: A Critical Reader.* Macmillan, 2013.

Falb, Daniel. "Epistemologies of Art in the Anthropocene," 302-317. In Behnke, Christoph, et. al., *Art in the Periphery of the Center.* Sternberg Press, 2015. https://www.academia.edu/13447828/Epistemologies_of_Art_in_the_Anthropocene

Ferretti, Federico. "Anarchism, geography and social art", in C. Kosuch (ed.) *Anarchist Avant-Garde,* Amsterdam, Brill, 2019, 235-259. https://brill.com/view/book/edcoll/9789004410428/BP000013.xml

Foster, John Bellamy, Hannah Holleman, and Brett Clark. "Imperialism in the Anthropocene." *Monthly Review* (July 1, 2019) https://monthlyreview.org/2019/07/01/imperialism-in-the-anthropocene/

Fourth National Climate Assessment. Volume II: Impacts, Risks, and Adaptation in the United States. Washington, D.C.: U.S. Global Change Research Program, 2018. https://nca2018.globalchange.gov/

Ghosh, Amitav. *The Great Derangement: Climate Change and the Unthinkable* (Chicago: University of Chicago Press, 2016), 1-84.

Gilman-Opalsky, Richard and Stevphen Shukaitis. *Riotous Epistemology.* Leipzig and Colchester: Journal of Aesthetics & Protest Press and Autonomedia. 2019.

Girardet, Herbert. "Sustainability is unhelpful: we need to think about regeneration." *The Guardian* (June 10, 2013). https://www.theguardian.com/sustainable-business/blog/sustainability-unhelpful-think-regeneration

Glacken, Clarence J. *Traces on the Rhodian Shore.* Berkeley: University of California, 1967.

Global Alliance for the Rights of Nature. *www.therightsofnature.com*

Global Commission on Adaptation. *Adapt Now: A Global Call for Leadership on Climate Resilience.* World Resources Institution.

Goodell, Jeff. *The Water Will Come: Rising Seas, Sinking Cities, and the Remaking of the Civilized World.* New York: Little Brown, 2017.

Gómez-Barris, Macarena. "Life Otherwise at the Sea's Edge." *Open Rivers* 13 (Spring 2019): 27-47.

Gordon, Uri. "Darkness Falls: Revisiting anarchist politics in the age of collapse." https://medium.com/@uri.gordon/darkness-falls-revisiting-anarchist-politics-in-the-age-of-collapse-ac334d5f003e

Green Anarchy Collective. "Green Anarchy." http://www.anarchism.net/schools_greenanarchy.htm

Gregory, Alice. "Accidental Killers: The challenge of living after you've caused another's death." *The New Yorker*, September 18, 2017: 28-32.

Guth, Christine M. *Hokusai's Great Wave: Biography of a Global Icon.* Honolulu: University of Hawaii Press, 2015.

Hall, Matthew. "Beyond the human: extending ecological anarchism." *Environmental Politics*, 20:3, (2011), 374-390. https://doi.org/10.1080/09644016.2011.573360

Hall, Stephanie. "The Great American Sea Serpent." *Folklife Today.* Library of Congress. August 8, 2016. https://blogs.loc.gov/folklife/2016/08/great-american-sea-serpent/

Harman, Greg. "Your brain on climate change: why the threat produces apathy, not action. *The Guardian.* November 10, 2014. https://www.theguardian.com/sustainable-business/2014/nov/10/brain-climate-change-science-psychology-environment-elections

Hayden, Patrick. "Gilles Deleuze and Naturalism: A Convergence with Ecological Theory and Politics, 23-45. In Bernd Herzogenrath, Ed. *An [Un]likely Alliance: Thinking Environment[s] with Deleuze/Guattari.* Newcastle upon Tyne, Cambridge Scholars Publishing, 2008.

Hedges, Chris. "The Coming Collapse." https://www.commondreams.org/views/2018/05/21/coming-collapse

Heilman, Kenneth M. and Russel S. Donda, "Neuroscience and Fundamentalism." *Tikkun.* July 2009. http://www.tikkun.org/magazine/tik0709/frontpage/neuroscience

Hickey, Amber. *A Guidebook of Alternative Nows.* http://joaap.org/press/alternative-nows_hickey.htm

Hickling, David. *The Ethical Significance of Levinas's Aesthetics.* Master of Research, Dept. of Philosophy, Macquarie University, 2014. https://www.academia.edu/36129883/The_ethical_significance_of_Levinass_aesthetics

Horkheimer, Max and Theodor Adorno. *Dialectic of Enlightenment,* edited by Mieke Bal and Hent deVries, translated by Edmund Jephcott. Stanford: Stanford University Press, 2002.

Hornborg, Alf. "How localization can solve climate change." *BBC Future.* September 6, 2019. https://www.bbc.com/future/article/20190905-how-localisation-can-solve-climate-change

Horvath, Katie. "Social Ecology: Radicalizing the Climate." *Roar* 9 (December 2019). https://roarmag.org/magazine/social-ecology-climate-movement

Huhn, Tom. "Adorno and Kant." *The Encyclopedia of Aesthetics, 1 (Oxford University Press, 1998): 29-32.* http://tomhuhn.com/includes/pdf/publications/articles/9_adorno_kant_A.pdf

Infoshop. "An Anarchist FAQ – J.2 What Is Direct Action? January 1, 2018. http://www.infoshop.org/an-anarchist-faq-j-2-what-is-direct-action/

Ingram, David. "Delivery dilemma: Americans are ordering more, but the U.S. can only handle so much." *NBC News* (December 23, 2019). https://www.nbcnews.com/tech/tech-news/delivery-dilemma-americans-are-ordering-more-u-s-can-only-n1106426

Intergovernmental Panel on Climate Change (IPCC). *Special Report on the Ocean and Cryosphere in a Changing Climate.* Geneva, 2019. https://www.ipcc.ch/srocc/home/

Intergovernmental Panel on Climate Change (IPCC) *Special Report: Global Warming of 1.5°C.* Geneva, 2018). https://www.ipcc.ch/sr15/

Intergovernmental Panel on Climate Change (IPCC). Reports. Geneva. https://www.ipcc.ch/reports/

International Situationists #8 (January 1963), "Ideologies, Classes and the Domination of Nature," In *Situationist International Anthology,* Ed. and Trans., Ken Knabb. Berkeley CA: Bureau of Public Secrets, 1981.

Ives, Lucy. "Orphans of Dickens: The social novel at the end of society." *The Baffler* 44 (March/April 2019), 25-34.

Jackson, Sue, "Compartmentalising Culture: The Articulation and Consideration of Indigenous Values in Water Resource Management," *Australian Geographer* 37.1 (2006): 19-31.

Jacobson, Gavin. "Our age of anxiety: Gavin Jacobson considers why doom-mongering is back in fashion." The Times Literary Supplement. February 6, 2019. https://www.the-tls.co.uk/articles/public/age-of-anxiety-fear/

Jagodzinski, Jan, Ed. *Interrogating the Anthropocene: Ecology, Aesthetics, Pedagogy and the Future in Question*. Cham, Switzerland: Palgrave MacMillan, 2018.

Jamail, Dahr. "Brace for Impact, as the Climate "End Game" Has Arrived." *Truthout*, September 25, 2018. https://truthout.org/articles/brace-for-impact-as-the-climate-end-game-has-arrived/

Jamail, Dahr and Barbara Cecil. "Rethink Activism in the Face of Catastrophic Biological Collapse." *Truthout*, March 4, 2019. https://truthout.org/articles/climate-collapse-is-on-the-horizon-we-must-act-anyway/

Janicka, Iwona. "Are these Bubbles Anarchist: Peter Sloterdijk's *Spherology* and the Question of Anarchism." *Anarchist Studies*. 24.1 (2016): 62-84.

Jenkins, Nicholas. "Running on the Waves: Pollock, Lowell, Bishop and the American Ocean," *Yale* Review 195, 2 (April 2007): 46-82.

Jochem, Eberhard, et. al. "Steps Toward a 2000-Watt Society: Developing a White Paper on Research and Development of Energy-efficient Technologies." Pre-study. December 16, 2002. https://www.ethz.ch/content/dam/ethz/special-interest/mtec/cepe/cepe-dam/documents/people/ejochem/Jochem_Steps_towards_a_2000_Watt_Society.pdf

Jochem, Eberhard, Ed. "Steps towards a sustainable development: A White Book for R8D of energy-efficient technologies." March, 2004. https://www.novatlantis.ch/wp-content/uploads/2014/10/Weissbuch.pdf

Johnson, Nathan. "SUVs are back, and they're spewing a boggling amount of carbon." *Grist*. October 18, 2019. https://grist.org/article/suvs-are-back-and-theyre-spewing-a-boggling-amount-of-carbon/

Johnstone, Caitlin. "Society is Made of Narrative. Realizing This Is Awakening from the Matrix." *The Medium* (August 21, 2018). https://medium.com/@caityjohnstone/society-is-made-of-narrative-realizing-this-is-awakening-from-the-matrix-787c7e2539ae

Jones, Nicola. "How to stop data centres from gobbling up the world's electricity." *Nature*, September 13, 2018. https://www.nature.com/articles/d41586-018-06610-y

Jones, P., 2006. "Stomping with the elephants: feminist principles for radical solidarity." In: S. Best and A.J. Nocella II, eds. *Igniting a revolution: voices in defence of the earth*. Edinburgh: AK Press, 319–334.

Jones, P., 2009. "Free as a bird: natural anarchism in action." In: R. Aster, et al., Eds. *Contemporary anarchist studies: an introductory anthology of anarchy in the academy*. New York: Routledge, 236–246.

Jones, Ryan T. "Running into Whales: The History of the North Pacific from Below the Wave," *The American Historical Review* 118.2 (2013): 249-377.

Jungwirth, Tomáš. "Finding Hope in Panic." *Green European Journal*. https://www.greeneuropeanjournal.eu/finding-hope-in-panic

Kamarck, Elaine. "The Challenging Politics of Climate Change." *The Brookings Institute*. September 23, 2019. https://www.brookings.edu/research/the-challenging-politics-of-climate-change

Kandelaars, Michael. "Marxism and the Natural World." *Marxist Left Review* 11 (Summer 2016). https://marxistleftreview.org/articles/marxism-and-the-natural-world

Kant, Immanuel. *Critique of Judgment*. Trans. Werner Pluhar, Indianapolis: Hackett, 1987.

Kaushall, J. Neville. *The Historical Negation of Aesthetic Categories: Adorno's Inheritance of Kant's Critique of Judgment*. PhD Thesis, University of Warwick, Department of Philosophy, 2019. https://www.academia.edu/39377949/The_Historical_Negation_of_Aesthetic_Categories_Adornos_Inheritance_of_Kants_Critique_of_Judgment

Kearney, Richard. *Dialogues with Contemporary Continental Thinkers: The Phenomenological Heritage*. Manchester: Manchester University Press, 1984.

Kempf, Herve. *How the Rich Are Destroying the Planet*. Cambridge: Green Books, 2008.

Khozin, Grigori. *Talking about the Future*. Moscow: Progress Publishers, 1988.

Kilby, Bill. "A Psychologist Explains Why People Don't Give a Shit About Climate Change." *Films for Action*. June 10, 2015. https://www.filmsforaction.org/articles/a-psychologist-explains-why-people-dont-give-a-shit-about-climate-change/

Kinna, Ruth. "Kropotkin's Theory of Mutual Aid in Historical Context." *International Review of Social History* 40 (1995): 259-283. http://www.psiche-natura.it/fileadmin/img/R._Kinna_Kropotkin_s_Theory_of_Mutual_Aid_in_Historical_Context.pdf

Klein, Naomi. *This Changes Everything: Capitalism vs. the Climate*. New York: Simon & Schuester, 2014.

Knabb, Ken. Ed. and Trans. Internationale Situationniste #8 (January 1963), "Ideologies, Classes and the Domination of Nature," 101-108. In *Situationist International Anthology*. Berkeley, CA: Bureau of Public Secrets, 1981, 1989.

Koehline, James. *Personal Correspondence*. December 2019.

Kolbert, Elizabeth. ""The Island in the Wind: A Danish Community's Victory over Carbon Emissions." *The New Yorker* (July 7 & 14) 2008. https://www.newyorker.com/magazine/2008/07/07

Kovel, Joel. *The Enemy of Nature: The End of Capitalism or the End of the World?* London and New York: Zed Books, 2007.

Kropotkin, Peter. *Mutual Aid: A Factor of Evolution*. Charleston, SC: Forgotten Books, 2008.

Kruse, Jamie and Elizabeth Ellsworth. "Design Spec in the Anthropocene: Imagining the Force of 30,000 Years of Geologic Change." In Heather Davis and Etienne Turpin, Eds. *Art in the Anthropocene: Encounters among Aesthetics, Politics,*

Environments and Epistemologies. Open Humanities Press, 2015óéé. http://openhumanitiespress.org/books/art-in-the-anthropocene

LaDuke, Winona. "The White Earth Band of Ojibwe Legally Recognized the Rights of Wild Rice. Here's Why." *Yes Magazine,* February 1, 2019. https://www.yesmagazine.org/planet/the-white-earth-band-of-ojibwe-legally-recognized-the-rights-of-wild-rice-heres-why-20190201

Lake Erie Bill of Rights. https://beyondpesticides.org/assets/media/documents/LakeErieBillofRights.pdf

Léger, Marc James. *Don't Network: The Avant Garde after Networks.* Colchester/New York/Port Townsend: Minor Compositions, 2018.

Leiss, William. *The Domination of Nature.* Boston: Beacon Press, 1972.

Leiss, William. *The Limits to Satisfaction: An Essay on the Problem of Needs and Commodities.* Toronto and Buffalo: University of Toronto Press, 1976.

LeMenager, Stephanie. *Living Oil: Petroleum Culture in the American Century.* Oxford & New York: Oxford University Press, 2014.

Levinas, Emmanuel. *Ethics and Infinity: Dialogues of Emmanuel Levinas and Philippe Nemo.* Pittsburgh: Duquesne University Press, 1995.

Lewis, Simon L. and Mark A. Maslin. *The Human Planet: How We Created the Anthropocene.* London: Yale University Press, 2018.

Lewis, Simon L. and Mark A. Maslin, "Defining the Anthropocene," Nature 519 (2015): 171-180.

Liboiron, Max. "The Perils of Ruin Porn: Slow Violence and the Ethics of Representation." *Discard Studies.* https://discardstudies.com/2015/03/23/the-perils-of-ruin-porn-slow-violence-and-the-ethics-of-representation/

Lieberman, Bruce. "A Brief Introduction to Climate Change and Transportation." *Yale Climate Connections.* September, 2019. https://www.yaleclimateconnections.org/2019/09/a-brief-introduction-to-climate-change-and-transportation

Lopez, Barry. *Of Wolves and Men.* New York: Scribners, 1978.

Lotringer, Sylvére in conversation with Heather Davis and Etienne Turpin. "The Last Political Scene." In Heather Davis and Etienne, Turpin, Eds. *Art in the Anthropocene: Encounters among Aesthetics, Politics, Environments and Epistemologies.* Open Humanities Press, 2015. http://openhumanitiespress.org/books/art-in-the-anthropocene

Lovino, Serenello. "The Reverse of the Sublime: Dilemmas (and Resources) of the Anthropocene Garden." *Transformations in Environment and Society.* 3 (2019). http://www.environmentandsociety.org/perspectives/2019/3/reverse-sublime-dilemmas-and-resources-anthropocene-garden

Lynch, Tommy. "Why Hope is Dangerous When it Comes to Climate Change." *Slate* July 2017. https://slate.com/technology/2017/07/why-climate-change-discussions-need-apocalyptic-thinking.html

Macfarlane, Robert. *Underland: A Deep Time Journey.* NY and Longdon: W. W. Norton & Co., 2019.

_____. "Should This Tree Have the Same Rights As You?" *The Guardian* (November 2, 2019). https://www.theguardian.com/books/2019/nov/02/trees-have-rights-too-robert-macfarlane-on-the-new-laws-of-nature

McGraw, Daniel. "Fighting pollution: Toledo residents want personhood status for Lake Erie." *The Guardian*. February 19, 2019. https://www.theguardian.com/us-news/2019/feb/19/lake-erie-pollution-personhood-status-toledo

MacKenzie, Donald A. *Myths of Babylonia and Assyria.* 1915. http://www.sacred-texts.com/ane/mba/mba13.htm

Malabou, Catherine. "The Brain of History or the Mentality of the Anthropocene." http://saq.dukejournals.org/content/116/1/39.abstract

Marlon, Jennifer R., B. Bloodhart, M. Ballew, J. Rolf-Reding, C. Roser-Renough, A. Leiserowitz and E. Mailbach. "How Hope and Doubt Affect Climate Change Mobilization." *Frontiers in Communication* 4 (May 2019) 1-15. https://www.frontiersin.org/articles/10.3389/fcomm.2019.00020/full

Marshall, George. (2014) *Don't Even Think About It: Why Our Brains Are Wired to Ignore Climate Change,* New York: Bloomsbury USA, 2014.

Martin, Elaine. "Re-reading Adorno: The 'after-Auschwitz' Aporia." *Forum.* University of Edinburgh, *Postgraduate Journal of Culture and the Arts* 02 | (Spring 2006). http://www.forumjournal.org/article/view/556

Martinko, Katherine. "The UN has declared war on ocean plastic pollution." *Treehugger.* February 27, 2017. http://www.treehugger.com/environmental-policy/un-says-its-time-tackle-plastic-pollution-aggressively.html

Marvel, Kate. "We Should Never Have Call It Earth." https://onbeing.org/blog/kate-marvel-we-should-never-have-called-it-earth/

Marx, Karl. *Economic and Philosophical Manuscripts of 1844, First Manuscript. Moscow: Progress Publishers, 1959.*

Mason, Paul. "The end of capitalism has begun." *The Guardian.* November 29, 2017. https://www.theguardian.com/books/2015/jul/17/postcapitalism-end-of-capitalism-begun

Mathieson, Charlotte (Ed.) *Sea Narratives: Cultural Responses to the Sea, 1600-Present.* Palgrave Macmillan. UK. http://www.palgrave.com/it/book/9781137581150

Matilsky, Barbara C. *Vanishing Ice: Alpine and Polar Landscapes in Art, 1755-2012.* Bellingham, WA: Whatcom Museum, 2013.

McDaniel, Carl N. "Can We Change, Will We Change? 216-228. In Carl N. McDaniel, *Wisdom for a Living Planet.* San Antonio: Trinity University Press, 2005.

McKibben, Bill. *Falter: Has the Human Game Begun to Play Itself Out?* New York: Henry Holt, 2019.

McKibben, Bill. "Life on a Shrinking Planet." *The New Yorker* November 26, 2018. https://www.newyorker.com/magazine/2018/11/26/how-extreme-weather-is-shrinking-the-planet

McKibben, Bill. "Climate politicking isn't working. We need climate civil disobedience." *Los Angeles Times* (Oct 09, 2018). https://www.latimes.com/opinion/op-ed/la-oe-mckibben-necessity-defense-bagley-minn-20181009-story.html

Meador, Ron. "New outlook on global warming: Best prepare for social collapse, and soon." *Minnpost,* October 15, 2018. https://www.minnpost.com/earth-journal/2018/10/new-outlook-on-global-warming-best-prepare-for-social-collapse-and-soon

Melville, Herman. *Moby-Dick.* 1851. Feedbooks. http://www.feedbooks.com

Meyer, Stephen M. *The End of the Wild.* Cambridge, London: MIT Press, 2006.

Miller, Asher. "Why Should We Even Bother?" *Post Carbon Institute.* December 29, 2014. http://www.postcarbon.org/why-should-we-even-bother/

Mirzoeff, Nicholas. "Visualizing the Anthropocene." *Public Culture* 26.2 (2014): 213-22.

Mirzoeff, Nicholas, "The Sea and the Land: Biopower and Visuality from Slavery to Katrina," *Culture, Theory and Critique* 50 (2009): 291.

Mitchell, William J. Thomas, Ed. *Landscape and Power.* Chicago: University of Chicago Press, 1994.

Mitchell, William J. Thomas. "Imperial Landscapes," in W.J.T. Mitchell, *Landscape and Power.* University of Chicago Press, 1994.

Monbiot, George. "Forget 'the environment:' we need new words to convey life's wonders." https://www.theguardian.com/commentisfree/2017/aug/09/forget-the-environment-new-words-lifes-wonders-language

Moore, Jason W. *Anthropocene or Capitalocene? Nature, History, and the Crisis of Capitalism.* Oakland, CA: PM Press, 2016.

Morales, Ricardo Levins. "Resilience: Another Name for Life." *Ricardo Levins Morales Art Studio* August 12, 2019. https://rlmartstudio.wordpress.com/2019/08/09/resilience-another-name-for-life

Morton, Timothy. *Hyperobjects: Philosophy and Ecology after the End of the World.* Minneapolis and London: University of Minnesota Press, 2013. http://massively-invisibleobjects.org/wp-content/uploads/2015/04/Hyperobjects.pdf

_____. *Realist Magic: Objects, Ontology, Causality.* Ann Arbor: University of Michigan Library, 2013. http://www.openhumanitiespress.org/books/titles/realist-magic/

_____. *Ecology without Nature: Rethinking Environmental Aesthetics.* Harvard University Press, 2009.

_____. *Humankind: Solidarity with Nonhuman People.* London and New York: Verso, 2017.

_____. *Being Ecological.* Cambridge, MA: The MIT Press, 2018.

Naess, Arne. *Ecology, Community and Lifestyle: Outline of an Ecosophy.* Trans. And revised by David Rothenberg. Cambridge University Press, 1989.

Nagelhout, Marah. "Nature and the 'Industry that Scorched It": Adorno and Anthropocene Aesthetics. *Symploke,* 24: 1-2, 2016: 121-135.

National Aeronautics and Space Administration (NASA) Global Climate Change, Vital Signs of the Planet. http://climate.nasa.gov/effects/ For more detailed information on impacts of temperature increases: Alan Buis, "A Degree of Concern: Why Global Temperatures Matter (Jun19, 2019). https://climate.nasa.gov/news/2878/a-degree-of-concern-why-global-temperatures-matter/

Necyk, Bradley and Daniel Harvey. "'Like Watching a Movie': Notes on the Possibilities of Art in the Anthropocene." In Jagodzinski, Jan, Ed. *Interrogating the Anthropocene: Ecology, Aesthetics, Pedagogy and the Future in Question.* Cham, Switzerland: Palgrave MacMillan, 2018: 237-251.

Neelu. "(Collective) Displacement: Solve for Y, When X Is Too. Tough?" *Neelu* March 2019. http://www.neelu.net/edge-science/collective-displacement-prelude-to-planet-hot-or-not/

Neelu. "The #GreenNewDeal – OTHER reasons why it might be a good idea." *Neelu* February 2019. http://www.neelu.net/future-tech/the-greennewdeal-other-reasons-why-it-might-be-a-good-idea/

O'Donohoe, Stephanie and Darach Turley. 'Dealing with death: art, mortality and the marketplace." In Brown, Stephen and Anthony Patterson, Eds. *Imagining Marketing: Art, aesthetics and the avant-garde.* New York and London: Routledge, 2000, 85-103.

Oudemans, A. C. (Anthonie Cornelis). *The Great Sea-Serpent: An historical and critical treatise. With the reports of 187 appearances...the suppositions and suggestions of scientific and non-scientific persons, and the author's conclusions.* With 82 illustrations. Published by the Author, 1892. https://archive.org/details/greatseaserpenth00oude

Owen, Connor. "Social Ecology and Aesthetic Criticism." *Studies in Arts and Humanities* (2, 2), 2016. http://sahjournal.com/index.php/sah/article/view/78

Pandora, Wikipedia. https://en.wikipedia.org/wiki/Pandora

Parry, Marton. "What would be the impacts of climate change assuming no, or some, or much emissions control and sequestration? http://www.scor-int.org/High-CO2_World/MartinParry.pdf

Pastore, C. Rev. Gary Kroll, *America's Ocean Wilderness: A Cultural History of Twentieth-Century Exploration.* http://www.history.ac.uk/reviews/review/926

Pelowski, Matthew and Fuminori Akiba . "A model of art perception, evaluation and emotion in transformative aesthetic experience." *New Ideas in Psychology* 29 (August 2011) 80–97. https://www.researchgate.net/publication/247091159_A_model_of_art_perception_evaluation_and_emotion_in_transformative_aesthetic_experience

Pente, Patti. "Slow Motion Electric Chiaroscuro: An Experiment in Glitch-Anthropo-Scenic Landscape Art." In Jagodzinski, Jan, Ed. *Interrogating the Anthropocene: Ecology, Aesthetics, Pedagogy and the Future in Question.* Cham, Switzerland: Palgrave MacMillan, 2018: 277-298.

People for the Ethical Treatment of Animals (PETA), "Cruelty on the Internet," https://www.peta.org/action/get-active-online/cruelty-internet/

Phelan, Jake. "Seascapes: tides of thought and being in Western perceptions of the sea." Goldsmiths Anthropology Research Papers Eds: Mao Mollona, Emma Tarlo, Frances Pine, Olivia Swift. Goldsmiths College, University of London, New Cross, London, 2007. https://www.gold.ac.uk/media/documents-by-section/departments/anthropology/garp/garp14.pdf

Pope Francis. *Laudato Si': On Care for Our Common Home.* Huntington, IN: Our Sunday Visitor, Inc., 2015.

Pope, Kristen. "The growing frequency of extreme weather dulls people's awareness of climate change impacts, researchers say." *Yale Climate Connections.* April 17, 2019. https://www.yaleclimateconnections.org/2019/04/normalizing-weather-extremes-dulls-concerns-for-warming/

Prange, Sebastian R. "Scholars and the Sea: A Historiography of the Indian Ocean," *History Compass* 6/5 (2005): 241.57. https://www.tandfonline.com/doi/abs/10.1080/09528820902840607

Preston, Christopher. "Forget the Anthropocene: We've Entered the Synthetic Age." *Singularity Hub* May 12, 2019. https://singularityhub.com/2019/05/12/forget-the-anthropocene-weve-entered-the-synthetic-age/

Priestland, David. "Anarchism Could Help Save the World." https://www.theguardian.com/books/2015/jul/03/anarchism-could-help-save-the-world

Price, Wayne. "A Green New Deal vs. Revolutionary Ecosocialism." *Black Rose Anarchist Federation.* January 4, 2019. http://blackrosefed.org/green-new-deal-ecosocialism

Raboteau, Emily. "Climate Signs." *The New York Review of Books.* https://www.nybooks.com/daily/2019/02/01/climate-signs

Rae, Jennifer H. *Art & The Anthropocene: Processes of responsiveness and communication in an era of environmental uncertainty.* Dissertation, School of Art, College of Design and Social Context, Melbourne, AU: RMIT University, 2015. https://www.academia.edu/28252634/Art_and_The_Anthropocene_processes_of_responsiveness_and_communication_in_an_era_of_environmental_uncertainty

Rancière, Jacques. *The Politics of Aesthetics.* Trans. Gabriel Rockhill. London & New York: Continuum International Publishing Group, 2004.

Ray, Gene. "Hits: From Trauma and the Sublime to Radical Critique." *Third Text.* 23, 2 (2009), 135-149. https://www.researchgate.net/publication/233107459_HITS_From_Trauma_and_the_Sublime_to_Radical_Critique

Read, Jason. "Anthropocene and Anthropogenesis: Philosophical Anthropology and the Ends of Man." *The South Atlantic Quarterly* 116:2, April 2017, 257-273.

Reclus, Élisée. *Man and Nature: The impact of human activity on physical geography/ Concerning the awareness of nature in modern society.* Sydney, AU: Jura Media, 1864, 1866, 1995. https://libcom.org/files/Reclus%20-%20Man%20and%20Nature.pdf

Reclus, Élisée. *The Earth and Its Inhabitants: The Universal Geography.* London: J.S. Virtue, 1894.

Reclus, Élisée. *The Universal Geography: the earth and its inhabitants.* Edited by E.G. Ravenstein. London: J S Virtue and Co, 1886-1894. (19 Volumes). https://en.wikisource.org/wiki/The_Earth_and_its_Inhabitants

Redford, Duncan, Ed. *Maritime History and Identity: The Sea and Culture in the Modern World.* London: I.B.Tauris & Co,Ltd, 2013.

Reed, Bill. "Shifting from 'sustainability' to regeneration." September 13, 2007. https://doi.org/10.1080/09613210701475753

Rees, Sara. *Tasting the Sea.* https://socialsoups.com/tag/tasting-the-sea/

Riding, Christine. "Shipwreck, Self-Preservation and the Sublime." https://www.tate.org.uk/art/research-publications/the-sublime/christine-riding-shipwreck-self-preservation-and-the-sublime-r1133015

Roald, Tone and Simo Køppe. "Sense and Subjectivity. Hidden Potentials in Psychological Aesthetics." *Journal of Theoretical and Philosophical Psychology* 35, 1 (2015): 20-34.

Roberts, Bill. "Production in View: Allan Sekula's Fish Story and the Thawing of Postmodernism." *Tate Papers* no. 18. Autumn 2012. www.tate.org.uk Accessed 10 February 2017. http://www.tate.org.uk/research/publications/tate-papers/18/production-in-view-allan-sekulas-fish-story-and-the-thawing-of-postmodernism

Robitzski, Dan. "Gene-Hacking Mosquitoes to Be Infertile Backfired Spectacularly." *Futurism* (September 16, 2019). https://futurism.com/the-byte/gene-hack-mosquitoes-backfiring

Romm, Joe. "More Worried About 'Global Warming' Than 'Climate Change.'" *Think Progress.* May 27, 2014. http://thinkprogress.org/climate/2014/05/27/3441787/yale-poll-global-warming-climate-change/

Rowe, J. Stan. "Ethics and the Sea." *Sea Wind* 2/4, Oct/Dec. 1988. 23-25. http://www.ecospherics.net/pages/RoEthSea.html

Royal Collection Trust, https://www.rct.uk/collection/912377/a-deluge

Sachs, Jeffrey D. "What's the path to deep decarbonization?" *World Economic Forum.* December 2, 2015. https://www.weforum.org/agenda/2015/12/whats-the-path-to-deep-decarbonization/

Sanders, Ed. *Investigative Poetry.* San Francisco/Woodstock NY: City Lights/Blake Route Press, 1976. http://woodstockjournal.com/pdf/InvestigativePoetry.pdf

Sauer-Thompson, Gary. Adorno: "The Shudder." October 29, 2010. http://www.sauer-thompson.com/archives/philosophy/2010/10/adorno-the-shud.html

Sax, Boris. "Storytelling and the 'Information Overload,'" On the Horizon 14, 3 (Fall 2006), 147-151.

Schama, Simon. *Landscape and Memory.* New York: A. A. Knopf, 1995.

Schertow, John Ahni. "The Yurok Nation just established the rights of the Klamath River." *Intercontinental Cry.* May 20, 2019. https://intercontinentalcry.org/the-yurok-nation-just-established-the-rights-of-the-klamath-river/

Schleuning, Neala. *Artpolitik: Social Anarchist Aesthetics in an Age of Fragmentation.* Wivenhoe/New York/Port Wateson: Minor Compositions, 2013.

Schlosser, Katesa. *Der Signalismus in der Kunst der Naturvolker: Biologisch-psychologische Gesetzlichkeiten in den Abweichungen von der Norm des Vorbildes,* Kiel, 1952.

Schopenhauer, Arthur. *The World as Will and Idea.* Trans. R. B. Haldane, M.A. and J. Kemp, M.A. Vol. I. Seventh Edition. London: Kegan Paul, Trench, Trübner & Co., 1909.

Schulz, Kathryn. "Polar Expressed: What if an ancient story about the Far North came true? *The New Yorker.* April 24, 2017: 88-95.

Schumacher, Ernst. F. *Small is Beautiful: Economics As If People Mattered.* NY: Harper Collins, 1976.

Scranton, Roy. "We Broke the World: Facing the fact of extinction." *The Baffler* 47 (September/October 2019): 86-93.

Secretariat of the Pink & Purple Polka-Dot Pyrate Parrot Party. "Cataclysm Catechism." *Neelu.* September, 2019. http://www.neelu.net/edge-science/climate-cataclysm-catechism-the-ultrafluorescent-left-mutant-version/

Sekula, Allan. *Fish Story.* Dusseldorf: Richter Verlag, 1995.

Sethness-Castro, Javier. *Imperiled Life: Revolution against Climate Catastrophe.* AK Press, 2012.

Sholtz, Janae. "Intervals of Resistance: Being True to the Earth in the Light of the Anthropocene." In Jagodzinski, Jan, Ed. *Interrogating the Anthropocene: Ecology, Aesthetics, Pedagogy and the Future in Question.* Cham, Switzerland: Palgrave MacMillan, 2018: 179-199.

Shorto, Russell. Review. Michael Pye, *The Edge of the World. New York Times,* June 5, 2015.

Shukaitis, Stevphen. *Imaginal Machines: Autonomy & Self-Organization in the Revolutions of Everyday Life.* London/NYC/Port Watson: Minor Compositions, 2009.

Sibley, Mulford Q. *Nature and Civilization: Some Implications for Politics.* Itasca, IL: F.E. Peacock Publishers, Inc. 1977.

Sidaway, James D., Richard J. White, Gerónimo Barrera de la Torre, Federico Ferretti, Nicholas Jon Crane, Shona Loong, Larry Knopp, Carrie Mott, Farhang Rouhani, Jonathan M. Smith & Simon Springer (2017) *The Anarchist Roots of*

Geography: Toward Spatial Emancipation, *The AAG Review of Books*, 5:4, 281-296, https://doi.org/10.1080/2325548X.2017.1366846

Sim, David. "World Water Day 2017: 60 Powerful Photos to Make You Think Twice About Leaving the Tap Running." *IB Times*. March 23, 2017. UK. http://www.ibtimes.co.uk/world-water-day-2017-60-powerful-photos-make-you-think-twice-about-leaving-tap-running-1612594#slideshow/1600521

Sloterdijk, Peter. *Spheres, Vol 1: Bubbles, Microspherology*. Trans. Wieland Hoban. South Pasadena, CA: Semiotext(E), 2011.

_____. *Spheres, Vol 2: Globes, Macrospherology*. Trans. Wieland Hoban. South Pasadena, CA: Semiotext(E), 2014.

_____. *Spheres, Vol 3: Foams, Plural Spherology*. Trans. Wieland Hoban. South Pasadena, CA: Semiotext(E), 2016.

_____. "The Anthropocene: A Process-State at the Edge of Geohistory?" In Heather Davis and Etienne Turpin, Eds. *Art in the Anthropocene: Encounters among Aesthetics, Politics, Environments and Epistemologies*. Open Humanities Press, 2015. http://openhumanitiespress.org/books/download/Davis-Turpin_2015_Art-in-the-Anthropocene.pdf

Smith, P.D. Rev. *The Sea: A Cultural History*. *Guardian*. September 20, 2013. https://www.theguardian.com/books/2013/sep/20/sea-cultural-history-john-mack-review

Smith, Richard. "An Ecosocialist Path to Limiting Global Temperature Rise to 1.5C." *System Change Not Climate Change, November 26, 2018*. https://systemchangenotclimatechange.org/article/ecosocialist-path-limiting-global-temperature-rise-15%C2%B0c

Smyth, Richard. "Nature writing's fascist roots." *New Statesman*. April 3, 2019. https://www.newstatesman.com/culture/books/2019/04/eco-facism-nature-writing-nazi-far-right-nostalgia-england

Sparrow, Rob. *Anarchist Politics & Direct Action*. https://theanarchistlibrary.org/library/rob-sparrow-anarchist-politics-direct-action#toc4

Springer, Simon. "Anarchist Geography." *The International Encyclopedia of Geography*. Edited by Douglas Richardson, Noel Castree, Michael F. Goodchild, Audrey Kobayashi, Weidong Liu, and Richard A. Marston. 2017 John Wiley & Sons, Ltd.

_____. "Anarchism and Geography: A Brief Genealogy of Anarchist Geographies." *Geography Compass* 7/1 (2013): 46–60. https://www.academia.edu/2006727/Anarchism_and_geography_a_brief_genealogy_of_anarchist_geographies

_____. "Total Liberation Ecology: Integral Anarchism, Anthroparchy, and the Violence of Indifference." https://www.academia.edu/38194539/Total_Liberation_Ecology_Integral_Anarchism_Anthroparchy_and_the_Violence_of_Indifference

_____. "Anarchist Geography." *The Wiley-AAG International Encyclopedia of Geography: People, the Earth, Environment, and Technology* (2015). https://www.academia.edu/6580928/Anarchist_geography

Spunk Library: an online anarchist library and archive. http://www.spunk.org/texts/intro/sp001695.html

Stabenow, Debbie - U. S. Senator, Ranking Member, Senate Committee on Agriculture, Nutrition and Forestry. "Peer-Reviewed Research on Climate Change by USDA Authors." January 2017-August 2019. https://www.politico.com/f/?id=0000016d-4aa1-de7e-ab6d-efb938460000

"Statement by President Trump on the Paris Climate Accord" (June 1, 2017). https://www.whitehouse.gov/briefings-statements/statement-president-trump-paris-climate-accord/

Stoknes, Per Espen. "The Great Grief: How To Cope with Losing Our World" *Common Dreams*. May 14, 2015. http://www.commondreams.org/views/2015/05/14/great-grief-how-cope-losing-our-world

Stone, Alison. "Adorno and the Disenchantment of Nature." *Philosophy and Social Criticism* 32 (2):231-253 (2006).

Strang, Veronica. "Fluid Consistencies: Material Relationality in Human Engagements with Water," *Archaeological Dialogues* 21.2 (2014), 133–150.

Sugimoto, Hiroshi. *Seascapes*. Bologna, IT, Damiani: 2015.

Swanson, Heather Anne. "The Banality of the Anthropocene." *Cultural Anthropology*. https://culanth.org/fieldsights/1074-the-banality-of-the-anthropocene

Talbot, Mary M. and Bryan Talbot. *Rain*. Milwaukie, OR: Dark Horse Comics, 2019.

Tallis, Raymond. "Art, humanity and the 'fourth hunger.'" *The Spiked Review of Books*. November 30, 2007. http://www.spiked-online.com/

Tangyin, Kajornpat. "Reading Levinas on Ethical Responsibility." https://www.academia.edu/606687/Reading_Levinas_on_Ethical_Responsibility

Taylor, NAJ. "The Falling man: 9/11's private moments." https://www.aljazeera.com/indepth/opinion/2011/09/201191014413515812.html

Thomas, Jack. "An apology from an environmentalist." *The Spinoff* July 1, 2019. https://thespinoff.co.nz/science/01-07-2019/an-apology-from-an-environmentalist

Thomassen, Bjørn. "Liminality" in *The Encyclopedia of Social Theory* (London 2006).

Tilghman, Joshua. "Leviathan the Sea Serpent: Our Emotional Energy Reservoir." The Spirit of the Scripture. January 18, 2013. http://www.spiritofthescripture.com/id1087-leviathan-the-sea-serpent-and-our-emotional-energy-reservoir.html

Tokar, Brian. "The liberatory potential of local action." *Roar*. September 19, 2019. https://roarmag.org/essays/liberatory-potential-local- action-tokar/

Toomey, Diane. "Climate Change and the Human Mind: A Noted Psychiatrist Weighs In." *Yale Environment 360*. October 26, 2017. https://e360.yale.edu/features/climate-change-and-the-human-mind-a-noted-psychiatrist-weighs-in

Truscello, Michael. "The New Topographics, Dark Ecology, and the Energy Infrastructure of Nations" *Imaginations* 3:2 (2012). https://journals.library.ualberta.ca/imaginations/index.php/imaginations/article/view/27253

_____. "Catastrophism and Its Critics: On the New Genre of Environmentalist Documentary Film." In Jagodzinski, Jan, Ed. *Interrogating the Anthropocene: Ecology, Aesthetics, Pedagogy and the Future in Question.* Cham, Switzerland: Palgrave MacMillan, 2018: 257-275.

Tsing, Anna Lowenhaupt. *The Mushroom at the End of the World: On the Possibility of Life in Capitalist Ruins.* Princeton, NJ: Princeton University Press, 2015.

Tsing, Anna, Heather Swanson, Elaine Gan, and Nils Bubandt, Eds. *Arts of Living on a Damaged Planet.* Minneapolis, MN: University of Minnesota Press, 2017.

Tuan, Yi-Fu. *Topophilia: A Study of Environmental Perception, Attitudes, and Values.* New York: Columbia University Press, 1974, 1990.

U.N. Environment. *Global Environment Outlook – GEO-6: Healthy Planet, Healthy People.* Cambridge University Press: August 2019. https://www.unenvironment.org/resources/global-environment-outlook-6

US Energy Information Administration. "Annual Energy Outlook 2020 with projections to 2050" (January 2020). https://www.eia.gov/outlooks/aeo/pdf/AEO2020%20Full%20Report.pdf

U. S. Environmental Protection Agency. "National Overview: Facts and Figures on Materials, Wastes and Recycling." https://www.epa.gov/facts-and-figures-about-materials-waste-and-recycling/national-overview-facts-and-figures-materials#Recycling/Composting

van Dooren, Thom. *The Wake of Crows: Living and Dying in Shared Worlds.* New York: Columbia University Press, 2019.

Vedantam, Shankar. "The Cassandra Curse: Why We Heed Sometime Warnings and Ignore Others." Radio Broadcast, National Public Radio. https://www.npr.org/2018/09/17/648781756/the-cassandra-curse-why-we-heed-some-warnings-and-ignore-others

Vernadsky, Vladimir. "The Biosphere and the Noosphere," *American Scientist* 33, no. 1 (1945) 1-12.

Vida V de Voss, *Emmanuel Levinas on Ethics as the First Truth.* Master of Arts Thesis, University of Stellenbosch, April, 2006. https://philpapers.org/rec/DEVELO

Wagner, Kate. "Staring at Hell: The aesthetic of architecture in a ruined world." *The Baffler* 49 (January-February, 2020): 88-101.

Wagoner, David. "Lost," In *Traveling Light: Collected and New Poems.* Champaign, IL: University of Illinois Press, 1999.

Wahl, Daniel C. "Sustainability is not enough: we need regenerative cultures." *Insurgintelligence.* March 15, 2017. https://medium.com/insurge-intelligence/sustainability-is-not-enough-we-need-regenerative-cultures-4abb3c78e68b

———. *Designing Regenerative Cultures*. Axminster, England: Triarchy Press, 2016.

Wallace-Wells, David. "Time to Panic." *New York Times* (February 16, 2019). https://www.nytimes.com/2019/02/16/opinion/sunday/fear-panic-climate-change-warming.html

———. *The Uninhabitable Earth: Life after Warming*. New York: Tim Duggan Books, 2019.

Wark, McKenzie. *Molecular Red: Theory for Anthropocene*. London: Verso, 2015.

Waters, Theodore E. A. and Robyn Fivush. "Relations between Narrative Coherence, Identity, and Psychological Well-being in Emerging Adulthood." https://www.ncbi.nlm.nih.gov/pmc/articles/PMC4324396/

"What's the path to deep decarbonization?" *World Economic Forum*, December 2, 2015. https://www.weforum.org/agenda/2015/12/whats-the-path-to-*deep*-decarbonization/

White, Damian Finbar and Gideon Kossoff. "Anarchism, Libertarianism and Environmentalism: Anti-Authoritarian Thought and the Search for Self-Organizing Societies," 50-65. In *The SAGE Handbook of Environment and Society*. Eds. Jules Pretty, Andy Ball, Ted Benton, Julia Guivant, David R Lee, David Orr, Max Pfeffer, Professor Hugh Ward. Los Angeles and Others, 2007.

Wildermuth, Ryhd. "The Future is Fascist." *Gods and Radicals Press* (February 28, 2019). https://abeautifulresistance.org/site/2019/2/28/jthe-future-is-fascist

Wilkinson, Jayne. "Liquid Economies, Networks of the Anthropocene." *Open Rivers* 3 (Summer, 2016): 14-24. http://editions.lib.umn.edu/openrivers/article/liquid-economies-networks-of-the-anthropocene/

Willis, Matthew. "Where Be Monsters? The *Daedalus* Sea Serpent and the War for Credibility." *The Appendix*, April 2014 2, 2. http://theappendix.net/issues/2014/4/where-be-monsters-daedalus-sea-serpent-and-war-for-credibility

Wilson, Edward O. *Half-Earth: Our Planet's Fight for Life*. New York and London: Liveright, 2016.

Wilson, Mark. "Smartphones Are Killing the Planet Faster Than Anyone Expected." *Fast Company*. March 3, 2018. https://www.fastcompany.com/90165365/smartphones-are-wrecking-the-planet-faster-than-anyone-expected

Woodhouse, Keith Makoto. *The Ecocentrists: A History of Radical Environmentalism*. New York: Columbia University Press, 2018.

Yamashita, Karen Tei. *I Hotel*. Minneapolis: Coffee House Press, 2010.

Yusoff, Kathryn and Jennifer Gabrys. "Climate change and the Imagination." *Wiley Interdisciplinary Reviews: Climate Change*, 2, 4, July, 2011, 516–534.

Zerzan, John. "The Case against Art." In Elements of Refusal, by John Zerzan, 54-62. Seattle: Left Bank Books, 1988.

Zerzan, John. "The Prison of Symbols." In Victoria Brunetta and Kate O'Shea, Eds. *Durty Words: A space for dialogue, solidarity, resistance and creation.* Limerick, Ireland: Durty Books, 2019. 210-213.

Zuidervaart, Lambert. "Theodor W. Adorno: Exposing capitalism's blind domination." *The Times Literary Supplement.* August 28, 2019. https://www.the-tls.co.uk/articles/public/theodor-adorno-footnotes-to-plato

ENDNOTES

Introduction

1. Pandora, Wikipedia. https://en.wikipedia.org/wiki/Pandora
2. Dipesh Chakrabarty, "The Climate of History: Four Theses," *Critical Inquiry* 35, 2 (Winter 2009), 197-222, 206-207.
3. Intergovernmental Panel on Climate Change (IPCC) *Special Report: Global Warming of 1.5°C* Geneva, 2018). https://www.ipcc.ch/sr15/
4. John Cook, et. al., "Consensus on consensus: a synthesis of consensus estimates on human-caused global warming" Environmental Research Letters, 2016. https://iopscience.iop.org/article/10.1088/1748-9326/11/4/048002/pdf
5. "Statement by President Trump on the Paris Climate Accord" (June 1, 2017). https://www.whitehouse.gov/briefings-statements/statement-president-trump-paris-climate-accord/
6. Earth System Research Monitoring Laboratory, Global Monitoring Division. "History of atmospheric carbon dioxide from 800,000 years ago until January, 2019. https://www.esrl.noaa.gov/gmd/ccgg/trends/history.html
7. NASA Global Climate Change, Vital Signs of the Planet. http://climate.nasa.gov/effects/ and for more detailed information on impacts of temperature increases: Alan Buis, "A Degree of Concern: Why Global Temperatures Matter (Jun19, 2019). https://climate.nasa.gov/news/2878/a-degree-of-concern-why-global-temperatures-matter/
8. David Wallace-Wells, *The Uninhabitable Earth: Life after Warming* (New York: Tim Duggan Books, 2019), 45.
9. Renee Cho, "Climate Change Poses Challenges to Plants and Animals." Earth Institute Columbia University (February 3, 2015). https://blogs.ei.columbia.edu/2015/02/03/climate-change-poses-challenges-to-plants-and-animals/

10　David Biello, "Mass Extinctions Tied to Past Climate Changes" *Scientific American* (October 24, 2007). https://www.scientificamerican.com/article/mass-extinctions-tied-to-past-climate-changes/

11　US Energy Information Administration. "Annual Energy Outlook 2020 with projections to 2050" (January 2020). https://www.eia.gov/outlooks/aeo/pdf/AEO2020%20Full%20Report.pdf

12　Martin Parry, "What would be the impacts of climate change assuming no, or some, or much emissions control and sequestration? http://www.scor-int.org/High-CO2_World/MartinParry.pdf?"

13　John Bellamy Foster, Hannah Holleman, and Brett Clark, "Imperialism in the Anthropocene," *Monthly Review* (July 1, 2019). https://monthlyreview.org/2019/07/01/imperialism-in-the-anthropocene/

14　"Does the USA have the highest cumulative CO_2 emissions since 1750?" *Carbon Brief: Clear on Climate.* 2018. https://skeptics.stackexchange.com/questions/45065/does-the-usa-have-the-highest-cumulative-co2-emissions-since-1750. Carbon Brief is a UK based website that tracks policy and scientific data on climate change. The video tracks global CO2 emissions from 1750 to the present. https://www.youtube.com/watch?v=jx85qK1ztAc.

15　Vladimir Vernadsky, "The Biosphere and the Noosphere," *American Scientist* 33, 1 (1945), 1-12.

16　Bill Kilby, "A Psychologist Explains Why People Don't Give a Shit About Climate Change," *Films for Action* (June 10, 2015). https://www.filmsforaction.org/articles/a-psychologist-explains-why-people-dont-give-a-shit-about-climate-change/

17　David Abram, *The Spell of the Sensuous: Perception and Language in a More-Than-Human World* (New York: Vintage, 1996, 2017): 266.

18　Thorstein Veblen, *The Theory of the Leisure Class: An Economic Study of Institutions* (NY: Macmillian, 1899).

19　Michael Truscello, Personal Communication.

20　George Monbiot, "Forget 'the environment:' we need new words to convey life's wonders," August 9, 2017. https://www.theguardian.com/commentisfree/2017/aug/09/forget-the-environment-new-words-lifes-wonders-language

21　Joe Romm, "More Worried About 'Global Warming' Than 'Climate Change,'" *Think Progress* (May 27, 2014). http://thinkprogress.org/climate/2014/05/27/3441787/yale-poll-global-warming-climate-change/

22　Timothy Morton, *Hyperobjects: Philosophy and Ecology after the End of the World* (Minneapolis and London: University of Minnesota Press, 2013), 15.

The Changing Sublime

1　Neala Schleuning, "There are Ten Thousand Grasses: Song of the Prairie." Manuscript.

2　Edmund Burke, *A Philosophical Enquiry into the Origin of Our Ideas of the Sublime and Beautiful,* Second edition (1759). http://www.english.upenn.edu/~mgamer/Etexts/burke-sublime.html

3   Edmund Burke, *Philosophical Enquiry into the Sublime and Beautiful and Other Pre-Revolutionary Writings*, Ed. David Womersley (London 1998), 102.
4   Christine Riding, "Shipwreck, Self-Preservation and the Sublime," *Tate*. https://www.tate.org.uk/art/research-publications/the-sublime/christine-riding-shipwreck-self-preservation-and-the-sublime-r1133015
5   Gene Ray, "Hits: From Trauma and the Sublime to Radical Critique," *Third Text* 23, 2 (2009), 135-149, 141. file:///C:/Users/river/Downloads/HITS_From_Trauma_and_the_Sublime_to_Radi.pdf
6   Arthur Schopenhauer, *The World as Will and Idea*, Trans. R. B. Haldane, M.A. and J. Kemp, M.A., Vol. I, Seventh Edition (London: Kegan Paul, Trench, Trübner & Co.), 1909, 271.
7   Max Horkheimer and Theodor Adorno, *Dialectic of Enlightenment*, Eds. Mieke Bal and Hent deVries, Trans. Edmund Jephcott (Stanford: Stanford University Press), 2002, 5-6.
8   Ibid, 31-32.
9   Ibid, 9.
10 Ibid, 19.
11 Theodor W. Adorno, *Aesthetic Theory*, Eds. Gretel Adorno and Rolf Tiedemann, Trans. with a translator's introduction by Robert Hullot-Kentor (Minneapolis: University of Minnesota Press,1997), 79-80.
12 Ibid., 244.
13 Marah Nagelhout, "Nature and the 'Industry that Scorched It: Adorno and Anthropocene Aesthetics," *Symploke* 24, 1-2 (2016), 121-135.
14 Ibid., 133.
15 Ibid., 121 and 132.
16 Ibid., 131.
17 Ibid., 133.
18 Timothy Morton, *Hyperobjects: Philosophy and Ecology after the End of the World* (Minneapolis and London: University of Minnesota Press, 2013), 17.
19 Ibid., 48.
20 Ibid., 36.
21 Ibid., 181.
22 Ibid., 130.
23 Ibid., 21.
24 Ibid., 132.
25 Ibid., 169.
26 Timothy Morton, *Realist Magic: Objects, Ontology, Causality* (Ann Arbor: University of Michigan Library, 2013), Chapter 2. http://www.openhumanitiespress.org/books/titles/realist-magic/
27 Emmanuel Levinas, *Ethics and Infinity: Dialogues of Emmanuel Levinas and Philippe Nemo* (Pittsburgh: Duquesne University Press, 1995), 95.

28 David Hickling, *The Ethical Significance of Levinas's Aesthetics*, Master of Research, Dept. of Philosophy, Macquarie University, 2014, 10. https://www.academia.edu/36129883/The_ethical_significance_of_Levinass_aesthetics

29 Kajornpat Tangyin, "Reading Levinas on Ethical Responsibility," *Responsibility and Commitment: Eighteen Essays in Honor of Gerhold K. Becker* (Edition Gorz, January 1, 2008), 155. https://www.academia.edu/606687/Reading_Levinas_on_Ethical_Responsibility

30 Richard. Kearney, *Dialogues with Contemporary Continental Thinkers: The Phenomenological Heritage* (Manchester: Manchester University Press, 1984), 63.

31 Tangyin, 157.

32 Peter Sloterdijk, *Spheres, Vol 3: Foams, Plural Spherology*, Trans. Wieland Hoban (South Pasadena, CA: Semiotext (E), 2016), 432.

33 Walt Whitman, "Song of Myself," *Leaves of Grass* (New York: Barnes and Nobles Books, 1992), 25-76, 42-43.

34 Heather Davis and Etienne, Turpin, "Art & Death: Lives between the Fifth Assessment & the Sixth Extinction," In Heather Davis and Etienne, Turpin, Eds. *Art in the Anthropocene: Encounters among Aesthetics, Politics, Environments and Epistemologies* (Open Humanities Press, 2015), 13. http://openhumanitiespress.org/books/art-in-the-anthropocene

35 Arnold Berleant, *Aesthetics and Environment: Theme and Variations on Art and Culture* (Ashgate Publishing: England and Vermont, 2005), 16.

36 Ibid., 150.

37 Morris Berman, *The Reenchantment of the World* (Ithaca: New York: Cornell University Press, 1981), 260.

38 Alison Stone, "Adorno and the Disenchantment of Nature," *Philosophy and Social Criticism* 32, 2(2006), 231-253, 232.

39 Ibid., 233.

40 Ibid., 243.

41 Ricardo Levins Morales, "Resilience: Another Name for Life," *Ricardo Levins Morales Art Studio* (August 12, 2019). https://rlmartstudio.wordpress.com/2019/08/09/resilience-another-name-for-life/?fbclid=IwAR1ZnNYFp6z9BtfNzEJ454jIjlXXtgoLU89kE2pbUDe-2vBjN8WjTaR6kjB0

42 Patti Pente, "Slow Motion Electric Chiaroscuro: An Experiment in Glitch-Anthropo-Scenic Landscape Art," In Jan Jagodzinski, Ed. *Interrogating the Anthropocene: Ecology, Aesthetics, Pedagogy and the Future in Question* (Cham, Switzerland: Palgrave MacMillan, 2018), 277-298, 284.

43 Ibid., 278.

44 Amitav Ghosh, *The Great Derangement: Climate Change and the Unthinkable* (Chicago: University of Chicago Press, 2016), 1-84, 30-31.

Making Art on a Dying Planet

1. Caitlin Johnstone, Society Is Made Of Narrative. Realizing This Is Awakening From The Matrix." *The Medium (August 21,2018).* https://medium.com/@caityjohnstone/society-is-made-of-narrative-realizing-this-is-awakening-from-the-matrix-787c7e2539ae
2. Xiangzhan Cheng, "On the Four Keystones of Ecological Aesthetic Appreciation," *East Asian Ecocriticism: A Critical Reader,* Eds., Simon C. Estok and Won-Chung Kim (Macmillan, 2013), 213-228, 228.
3. Jennifer H.Rae, *Art & The Anthropocene: Processes of responsiveness and communication in an era of environmental uncertainty,* Dissertation, School of Art, College of Design and Social Context, (Melbourne, AU: RMIT University, 2015), 78. https://www.academia.edu/28252634/Art_and_The_Anthropocene_processes_of_responsiveness_and_communication_in_an_era_of_environmental_uncertainty
4. Macarena Gómez-Barris, "Life Otherwise at the Sea's Edge," *Open Rivers* 13 (Spring 2019), 27-47, 29.
5. Tobias Boes and Kath Marshall, "Writing the Anthropocene: An Introduction," *Minnesota Review* 83 (2014), 60-72, 64.
6. Timothy Morton, *Hyperobjects: Philosophy and Ecology after the End of the World* (Minneapolis and London: University of Minnesota Press, 2013): 201.
7. Amitav Ghosh, *The Great Derangement: Climate Change and the Unthinkable* (Chicago: University of Chicago Press, 2016), 1-84, 83.
8. Ibid., 83 and 84.
9. Lucy Ives, "Orphans of Dickens: The social novel at the end of society," *The Baffler* 44 (March/April 2019), 25-34, 26.
10. David Bayles and Ted Orland, *Art & Fear: Observations on the Perils (and Rewards) of Artmaking* (Santa Cruz, CA and Eugene, OR: The Image Continuum, 1993), 108.
11. Lawrence Buell, *The Environmental Imagination: Thoreau, nature writing, and the formation of American Culture* (Cambridge, MA: Harvard University Press, 1995), 285.
12. Kathryn Yusoff and Jennifer Gabrys, "Climate change and the Imagination," *Wiley Interdisciplinary Reviews: Climate Change,* 2, 4 (July, 2011), 516–534, 521.
13. Timothy Morton, *Being Ecological* (Cambridge, MA: The MIT Press, 2018), xlii.
14. Yusoff and Gabrys, 520.
15. Ibid.
16. David Wallace-Wells, "Time to Panic," *New York Times* (February 16, 2019). https://www.nytimes.com/2019/02/16/opinion/sunday/fear-panic-climate-change-warming.html
17. Theodore Adorno, *Negative Dialectics,* Trans. E. B. Ashton (London and New York: Routledge, 1973, 2004), 369.
18. Sylvére Lotringer in conversation with Heather Davis and Etienne Turpin, "The Last Political Scene," In Heather Davis and Etienne, Turpin, Eds. *Art in the Anthropocene: Encounters among Aesthetics, Politics, Environments and Epistemologies* (Open Humanities Press, 2015), 373. http://openhumanitiespress.org/books/art-in-the-anthropocene
19. Peter Sloterdijk, "The Anthropocene: A Process-State at the Edge of Geohistory?" In Heather Davis and Etienne Turpin, Eds. *Art in the Anthropocene: Encounters among*

Aesthetics, Politics, Environments and Epistemologies. (Open Humanities Press, 2015), 330. http://openhumanitiespress.org/books/download/Davis-Turpin_2015_Art-in-the-Anthropocene.pdf

20   Ibid., 337.

21   Michael Truscello, "Catastrophism and Its Critics: On the New Genre of Environmentalist Documentary Film," In Jagodzinski, Jan, Ed. *Interrogating the Anthropocene: Ecology, Aesthetics, Pedagogy and the Future in Question.* (Cham, Switzerland: Palgrave MacMillan), 2018, 257-275, 258.

22   Christine M.Guth, *Hokusai's Great Wave: Biography of a Global Icon* (Honolulu: University of Hawaii Press, 2015), 206.

23   Wallace-Wells, 157.

24   Tomáš Jungwirth, "Finding Hope in Panic," *Green European Journal.* https://www.greeneuropeanjournal.eu/finding-hope-in-panic/?fbclid=IwAR2fURq96PYaDKw8jHohkg-Zpp7GrMRkegWBGD2ncJBnrk3rHUzudyfvD0q8

25   Gene Ray, "Hits: From Trauma and the Sublime to Radical Critique," *Third Text* 23, 2 (2009), 135-149, 138.

26   Ibid., 144.

27   NAJ Taylor, "The Falling man: 9/11's private moments." https://www.aljazeera.com/indepth/opinion/2011/09/201191014413515812.html

28   People for the Ethical Treatment of Animals (PETA), "Cruelty on the Internet," https://www.peta.org/action/get-active-online/cruelty-internet/

29   Michael Truscello, "The New Topographics, Dark Ecology, and the Energy Infrastructure of Nations," *Imaginations* 3:2 (2012), 199. https://journals.library.ualberta.ca/imaginations/index.php/imaginations/article/view/27253199.

30   Serenello Lovino, "The Reverse of the Sublime: Dilemmas (and Resources) of the Anthropocene Garden." *Transformations in Environment and Society* 3 (2019), 9. http://www.environmentandsociety.org/perspectives/2019/3/reverse-sublime-dilemmas-and-resources-anthropocene-garden

31   Ibid., 19.

32   Ibid., 22.

33   Ibid., 7.

34   Jan Jagodzinski, Ed., *Interrogating the Anthropocene: Ecology, Aesthetics, Pedagogy and the Future in Question* (Cham, Switzerland: Palgrave MacMillan, 2018), 45.

35   Kristen Pope, "The growing frequency of extreme weather dulls people's awareness of climate change impacts, researchers say," *Yale Climate Connections* (April 17, 2019). https://www.yaleclimateconnections.org/2019/04/normalizing-weather-extremes-dulls-concerns-for-warming/

36   Guy Debord, *Society of the Spectacle*, Trans. and Annotated by Ken Knabb (Berkeley, CA: Bureau of Public Secrets, 2014): Para. 4.

37   Guy Debord, *Comments on the Society of the Spectacle* (Pirate Press, 1991; Verso, 1990), 27-28.

38 International Situationists #8 (January 1963), "Ideologies, Classes and the Domination of Nature," In *Situationist International Anthology,* Ed. and Trans., Ken Knabb (Berkeley CA: Bureau of Public Secrets, 1981), 105.
39 Guy Debord, *Society*, Para. 203.
40 Stephanie LeMenager, *Living Oil: Petroleum Culture in the American Century* (Oxford & New York: Oxford University Press, 2014), 66.
41 Ibid., 67.
42 Ibid., 68.
43 Ibid., 70.
44 Ibid., 80 and 89.
45 Ibid., 92.
46 Ibid., 101.
47 Ibid., 104.
48 Brett Bloom, *Petro-subjectivity: De-Industrializing Our Sense of Self* (Aubun, IN: Breakdown Break Down Press, 2015, 2018), 22.
49 Jacques Rancière, *The Politics of Aesthetics,* Trans. Gabriel Rockhill (London & New York: Continuum International Publishing Group, 2004), 39.
50 Ibid., 63.
51 Boris Sax, "Storytelling and the 'Information Overload,'" *On the Horizon* 14, 3 (Fall 2006), 165-170, 167. https://www.academia.edu/265396/Storytelling_and_the_Information_Overload
52 Rae, 81.
53 Stevphen Shukaitis, *Imaginal Machines: Autonomy & Self-Organization in the Revolutions of Everyday Life* (London/New York/Port Watson: Minor Compositions, 2009), 107.
54 Raboteau, Emily. "Climate Signs." *The New York Review of Books.* https://www.nybooks.com/daily/2019/02/01/climate-signs/?fbclid=IwAR1xHflKOiA9JQiVB1Rr0CWKRRKRsVw-wcvHpXHDYFmhSumsVfaoE4ggU7kw
55 Matthew Hall, "Beyond the human: extending ecological anarchism," *Environmental Politics*, 20:3, (2011), 374-390, 387. https://doi.org/10.1080/09644016.2011.573360
56 Buell, 260.
57 Morton, *Being Ecological,* 72-73.
58 Arnold Berleant, *Aesthetics and Environment: Theme and Variations on Art and Culture* (Ashgate Publishing: England and Vermont, 2005), 81.
59 Jagodinski, 45.
60 Robert Yount, Personal Correspondence.
61 Diane Toomey, "Climate Change and the Human Mind: A Noted Psychiatrist Weighs In," *Yale Environment* (October 26, 2017). https://e360.yale.edu/features/climate-change-and-the-human-mind-a-noted-psychiatrist-weighs-in
62 McDaniel, 228.
63 Bill Reed, "Shifting from 'sustainability' to regeneration," *Building Research & Information* 35 (September 13, 2007), 7. https://doi.org/10.1080/09613210701475753
64 LeManager, 34-39.

65 Ibid., 44.
66 Claire Colebrook, "Earth Felt the Wound: The Affective Divide," *Journal for Politics, Gender, and Culture*, 8, 1, (2011), 45-58, 45.
67 Ibid., 55.
68 Janae Sholtz, "Intervals of Resistance: Being True to the Earth in the Light of the Anthropocene," In Jagodzinski, Jan, Ed., *Interrogating the Anthropocene: Ecology, Aesthetics, Pedagogy and the Future in Question* (Cham, Switzerland: Palgrave MacMillan, 2018), 179-199, 186.
69 William J. Brady, A. P. Gantman, and J. J. Van Bavel, "Attentional Capture Helps Explain Why Moral and Emotional Content Go Viral," *Journal of Experimental Psychology: General* (Advance online publication (September 5, 2019), 9. http://dx.doi.org/10.1037/xge0000673
70 Critical Art Ensemble, *Aesthetic, Necropolitics, and Environmental Struggle* (New York: Autonomedia, 2018), 67.
71 Ibid., 69-70.
72 Scholtz, 192.
73 Timothy Morton, *Hyperobjects: Philosophy and Ecology (after the End of the World* (Minneapolis and London: University of Minnesota Press, 2013), 183.
74 James Koehline, Personal Correspondence (December 2019).
75 Jennifer R. Marlon, B. Bloodhart, M. Ballew, J. Rolf-Reding, C. Roser-Renough, A. Leiserowitz and E. Mailbach "How Hope and Doubt Affect Climate Change Mobilization," *Frontiers in Communication* 4 (May 2019), 1-15, 1 and 2. https://www.frontiersin.org/articles/10.3389/fcomm.2019.00020/full
76 Ibid., 12.
77 Jem Bendell, "Deep Adaptation: A Map for Navigating Climate Tragedy," *IFLAS Occasional Paper* 2 (July 27, 2018). www.iflas.info
78 Ibid.

Do You See What I Sea?

1 Brett Bloom, *Petro-subjectivity: De-Industrializing Our Sense of Self* (Aubun, IN: Breakdown Break Down Press, 2015, 2018).
2 Royal Collection Trust.
3 Amitav Ghosh, *The Great Derangement: Climate Change and the Unthinkable* (Chicago: University of Chicago Press, 2016), 1-84, 55 and 37.
4 Ghosh, 37.
5 Jake Phelan, "Seascapes: tides of thought and being in Western perceptions of the sea," *Goldsmiths Anthropology Research Papers,* Eds., Mao Mollona, Emma Tarlo, Frances Pine, Olivia Swift (Goldsmiths College, University of London, New Cross, London, 2007), 10. https://www.gold.ac.uk/media/documents-by-section/departments/anthropology/garp/garp14.pdf

6 Nicholas Jenkins, "Running on the Waves: Pollock, Lowell, Bishop and the American Ocean," *Yale Review* 195, 2 (April 2007), 46-82, 56.
7 Jenkins, 49.
8 Jenkins, 58-59.
9 Jenkins, 80.
10 Allan Sekula, *Fish Story* (Dusseldorf: Richter Verlag, 1995).
11 Nicholas Mirzoeff, "The Sea and the Land: Biopower and Visuality from Slavery to Katrina," *Culture, Theory and Critique* 50 (2009), 291, 3.
12 Phelan, 7.
13 *Chaoskampf*, Wikipedia.
14 Herman Melville, *Moby-Dick* (Feedbooks, 1851), 309. http://www.feedbooks.com
15 William J. Broad, "The Ice Was Only Part of It" (*New York Times,* April 9, 2012). https://www.nytimes.com/2012/04/10/science/a-new-look-at-natures-role-in-the-titanics-sinking.html
16 Veronica Strang, "Fluid Consistencies: Material Relationality in Human Engagements with Water," *Archaeological Dialogues* 21, 2 (2014), 133–150.
17 Phelan, 1.
18 Jamie Kruse and Elizabeth Ellsworth, "Design Spec in the Anthropocene: Imagining the Force of 30,000 Years of Geologic Change," In Heather Davis and Etienne Turpin, Eds., *Art in the Anthropocene: Encounters among Aesthetics, Politics, Environments and Epistemologies* (Open Humanities Press, 2015). http://openhumanitiespress.org/books/art-in-the-anthropocene
19 Sue Jackson, "Compartmentalising Culture: The Articulation and Consideration of Indigenous Values in Water Resource Management," *Australian Geographer* 37, 1 (2006), 19-31 and Strang.
20 Phelan, 11.
21 Christine M. Guth, *Hokusai's Great Wave: Biography of a Global Icon* (Honolulu: University of Hawaii Press, 2015), 19.
22 Guth, 204.
23 Guth, 7.
24 Guth, 168.
25 Phelen, 11.
26 Melville, 255.
27 Mirzoeff, "Sea and the Land," 6.
28 Melville, 308.
29 Ibid., 309.
30 Alan Sekula, *Fish Story* (Dusseldorf: Richter Verlag, 1995).
31 Nicholas Mirzoeff, "Visualizing the Anthropocene," *Public Culture* 26, 2 (2014), 213-22, 213.
32 George Monbiot, "Forget 'the environment:' we need new words to convey life's wonders," *The Guardian* (August 9, 2017). https://www.theguardian.com/commentisfree/2017/aug/09/forget-the-environment-new-words-lifes-wonders-language

33 Mirzoeff, "Visualizing," 217.
34 Ghosh, 55.
35 Mirzoeff, "Visualizing," 217
36 Ghosh, 22 and 25.
37 Ghosh, 32.
38 Arnold Berleant, *Aesthetics and Environment: Theme and Variations on Art and Culture* (Ashgate Publishing: England and Vermont, 2005), 83.
39 Berleant, 98.

Nothing Will Ever Be the Same Again – Question Politics as Usual

1   David Wallace-Wells, *The Uninhabitable Earth: Life after Warming* (New York: Tim Duggan Books, 2019), 43.
2   Dipesh Chakrabarty, "The Climate of History: Four Theses," *Critical Inquiry* 35 (2) (Winter 2009), 197-222, 13.
3   Heather Davis and Etienne Turpin, "Art & Death: Lives between the Fifth Assessment & the Sixth Extinction," In Heather Davis and Etienne Turpin, Eds. *Art in the Anthropocene: Encounters among Aesthetics, Politics, Environments and Epistemologies* (Open Humanities Press, 2015), 7. http://openhumanitiespress.org/books/art-in-the-anthropocene
4   Simon L. Lewis, and Mark A. Maslin, *The Human Planet: How We Created the Anthropocene* London: Yale University Press, 2018), 369-371.
5   Timothy Morton, *Ecology without Nature: Rethinking Environmental Aesthetics* (Harvard University Press, 2009), 204.
6   Ibid., 185.
7   Alice Gregory, "Accidental Killers: The challenge of living after you've caused another's death," *The New Yorker* (September 18, 2017), 28-32, 32.
8   Nathan Jun, Personal Correspondence, June 2019.
9   Jem Bendell, "Deep Adaptation: A Map for Navigating Climate Tragedy," *IFLAS Occasional Paper* 2 (July 27, 2018). www.iflas.info
10 Kathryn Yusoff and Jennifer Gabrys, "Climate change and the Imagination," *Wiley Interdisciplinary Reviews: Climate Change*, 2, 4 (July, 2011), 516–534, 529.
11 Michael Truscello, "Catastrophism and Its Critics: On the New Genre of Environmentalist Documentary Film," In Jagodzinski, Jan, Ed., *Interrogating the Anthropocene: Ecology, Aesthetics, Pedagogy and the Future in Question* (Cham, Switzerland: Palgrave MacMillan, 2018), 257-275, 269.
12 Timothy Morton, *Being Ecological* (Cambridge, MA: The MIT Press), 2018, 153.
13 Wayne Price, "A Green New Deal vs. Revolutionary Ecosocialism," *Black Rose Anarchist Federation* (January 4, 2019). http://blackrosefed.org/green-new-deal-ecosocialism/?fbclid=IwAR0syRHEtEVHXex3MxulBM7XWp1-3bY4BMHLBud4jN7Ed78MqZd0Ww-WqUOI

14 Jeffrey D. Sachs, "What's the path to deep decarbonization?" *World Economic Forum* (December 2, 2015). https://www.weforum.org/agenda/2015/12/whats-the-path-to-deep-decarbonization/
15 Lewis and Maslin, 395-399.
16 P.D. Smith, Rev., *The Sea: A Cultural History, Guardian* (September 20, 2013). https://www.theguardian.com/books/2013/sep/20/sea-cultural-history-john-mack-review
17 U. S. Environmental Protection Agency, "National Overview: Facts and Figures on Materials, Wastes and Recycling," https://www.epa.gov/facts-and-figures-about-materials-waste-and-recycling/national-overview-facts-and-figures-materials#Recycling/Composting
18 Collaborative on Health and the Environment, "Precautionary Principle: The Wingspread Statement." https://www.healthandenvironment.org/environmental-health/social-context/history/precautionary-principle-the-wingspread-statement
19 Wallace-Wells, 227.
20 Ibid., 192-193.
21 Uri Gordon, "Darkness Falls: Revisiting anarchist politics in the age of collapse," *Medium* (August 9, 2018). https://medium.com/@uri.gordon/darkness-falls-revisiting-anarchist-politics-in-the-age-of-collapse-ac334d5f003e
22 Chris Hedges, "The Coming Collapse," *Common Dreams* (May 21, 2018) https://www.commondreams.org/views/2018/05/21/coming-collapse?utm_campaign=shareaholic&utm_medium=facebook&utm_source=socialnetwork&fbclid=IwAR1V-nh9wAEwM-G0utYT-VB3TdBHkW2Xbpb3oSHVefVBeVKpGSZ_WdCJHqjs
23 Bendell.
24 Ibid.
25 Dahr Jamail and Barbara Cecil. "Rethink Activism in the Face of Catastrophic Biological Collapse." *Truthout*, March 4, 2019. https://truthout.org/articles/climate-collapse-is-on-the-horizon-we-must-act-anyway/
26 Eberhard Jochem, Ed., "Steps towards a sustainable development: A White Book for R&D of energy-efficient technologies," *Novatlantis* (March, 2004). https://www.novatlantis.ch/wp-content/uploads/2014/10/Weissbuch.pdf
27 Paul Mason, "The end of capitalism has begun," *The Guardian* (November 29, 2017). https://www.theguardian.com/books/2015/jul/17/postcapitalism-end-of-capitalism-begun
28 Ibid.
29 Lewis and Maslin, 400-414.
30 In recent years, her research activities have expanded and will be available at http://anthropocene.au.dk/feral-atlas/.
31 Ryhd Wildermuth, "The Future is Fascist," *Gods and Radicals Press* (February 28, 2019). https://abeautifulresistance.org/site/2019/2/28/jthe-future-is-fascist?fbclid=IwAR032Tg-DW8lLKkmsXY7YMuTvvNhAbqJlfv6ciGpuaTEfRc8n6oPlpXo9SEc
32 Richard Smyth, "Nature writing's fascist roots," *New Statesman* (April 3, 2019). https://www.newstatesman.com/culture/books/2019/04/eco-facism-nature-writing

-nazi-far-right-nostalgia-england?fbclid=IwAR0dJFXOXm5Vq0sDJWSyNFLT-3LHRpc0sh_ZptwfWhswqVV7M7ZSe_rcbq8A

33 Christopher Preston, "Forget the Anthropocene: We've entered the Synthetic Age," *Singularity Hub* (May 12, 2019). https://singularityhub.com/2019/05/12/forget-the-anthropocene-weve-entered-the-synthetic-age/
34 Wallace-Wells, 171.
35 Brett Bloom, *Petro-subjectivity: De-Industrializing Our Sense of Self* (Aubun, IN: Breakdown Break Down Press, 2015, 2018), 49.
36 Ibid., 56.
37 Wallace-Wells, 179.
38 Ibid., 180.
39 Bendell.
40 Dan Robitzski. "Gene-Hacking Mosquitoes to Be Infertile Backfired Spectacularly." *Futurism* September 16, 2019. https://futurism.com/the-byte/gene-hack-mosquitoes-backfiring
41 Jasper Berns, "Between the Devil and the Green New Deal," *Commune* 3 (Summer 2019). https://communemag.com/between-the-devil-and-the-green-new-deal/?fbclid=IwAR06VFpAUQMMb7RJP4N6X9n-_8Ynx8kGF3l6wVnIhhExEVkKyeS5F3h9r7I
42 For a detailed analysis of projected energy use, see Nicola Jones, "How to stop data cenres from gobbling up the world's electricity," *Nature* (September 12, 2018). https://www.nature.com/articles/d41586-018-06610-y
43 Mulford Q. Sibley, *Nature and Civilization: Some Implications for Politics* (Itasca, IL: F.E. Peacock Publishers, Inc. 1977).
44 Jason Read, "Anthropocene and Anthropogenesis: Philosophical Anthropology and the Ends of Man," *The South Atlantic Quarterly* 116, 2 (April 2017), 257-273, 270-271.

Nothing Will Ever Be the Same Again - Anarchist Practice for the Anthropocene

1 Elaine Kamarck, "The Challenging Politics of Climate Change," *The Brookings Institute* (September 23, 2019). https://www.brookings.edu/research/the-challenging-politics-of-climate-change/?utm_campaign=Brookings%20Brief&utm_source=hs_email&utm_medium=email&utm_content=77189249
2 Daniel C. Wahl, *Designing Regenerative Cultures* (Axminster, England: Triarchy Press, 2016), 195.
3 Nathan Jun, Personal Correspondence, October, 2019
4 Richard Gilman-Opalsky and Stevphen Shukaitis, *Riotous Epistemology* (*Journal of Aesthetics & Protest Press* and *Autonomedia*, 2019), 58-59.
5 Common Ground Collective. Wikipedia.
6 Herve Kempf, *How the Rich Are Destroying the Planet* (Cambridge: Green Books), 2008.
7 Karl Marx, *Economic and Philosophical Manuscripts of 1844, First Manuscript (Moscow: Progress Publishers, 1959.*

8   Friedrich Engels, "The Part played by Labour in the Transition from Ape to Man," Trans. Clemens Dutt, 1876. https://www.marxists.org/archive/marx/works/1876/part-played-labour/index.htm (Progress Publishers, Moscow, 1934-English).
9   Matthew Hall, "Beyond the human: extending ecological anarchism," *Environmental Politics*, 20, 3 (2011), 374-390. https://doi.org/10.1080/09644016.2011.573360
10 John P. Clark, "The Dialectical Social Geography of Elisée Reclus," From *Philosophy and Geography I: Space, Place, and Environmental Ethics* (Lanham, MD: Rowman and Littlefield, 1997), 117-142: 2. https://www.academia.edu/2540498/_The_Dialectical_Social_Geography_of_Elisee_Reclus_
11 Ibid., 7.
12 Élisée Reclus, "Impact," *Man and Nature: The impact of human activity on physical geography/Concerning the awareness of nature in modern society* (Sydney, AU: Jura Media, 1864, 1866, 1995), 8. https://libcom.org/files/Reclus%20-%20Man%20and%20Nature.pdf
13 Ibid., "Awareness," 34.
14 Clark, 11.
15 Hall, 378.
16 Ruth Kinna, "Kropotkin's Theory of Mutual Aid in Historical Context," *International Review of Social History* 40 (1995), 259-283. http://www.psichenatura.it/fileadmin/img/R._Kinna_Kropotkin_s_Theory_of_Mutual_Aid_in_Historical_Context.pdf
17 Arne Naess, *Ecology, Community and Lifestyle: Outline of an Ecosophy*, Trans. And revised by David Rothenberg (Cambridge University Press, 1989).
18 Ernst. F. Schumacher, *Small is Beautiful: Economics As If People Mattered* (New York: Harper Collins, 1976), 69.
19 Ibid., 64.
20 Keith Makoto Woodhouse, *The Ecocentrists: A History of Radical Environmentalism* (New York: Columbia University Press, 2018), 286 and 288.
21 Ibid., 3.
22 Ibid., 71.
23 Ibid., 7.
24 Ibid., 129.
25 Ibid., 161.
26 Ibid., 51.
27 Katie Horvath, "Social Ecology: Radicalizing the Climate." *Roar* 9 (December 2019). https://roarmag.org/magazine/social-ecology-climate-movement/?fbclid=IwAR2YNO-41zUfgcZyjriJuBSNOtUIjxP6y_dQ4FW2YECXNrNlQ7-TDufg9RDk
28 Green Anarchy Collective, "Green Anarchy." http://www.anarchism.net/schools_greenanarchy.htm
29 Timothy Morton, *Humankind: Solidarity with Nonhuman People* (London and New York: Verso, 2017), 36-37.
30 Ibid., 133.
31 The Rights of Nature. therightsofnature.org

32 Winona LaDuke, "The White Earth Band of Ojibwe Legally Recognized the Rights of Wild Rice. Here's Why," *Yes Magazine* (February 1, 2019). https://www.yesmagazine.org/planet/the-white-earth-band-of-ojibwe-legally-recognized-the-rights-of-wild-rice-heres-why-20190201?fbclid=IwAR1z5wVWFl4k9BOaI5xpqjdtW90XCL7wJVBnoqmzW-Gr2TBPr6ag-ftjevC4

33 John Ahni Schertow, "The Yurok Nation just established the rights of the Klamath River," *Intercontinental Cry.* (May 20, 2019). https://intercontinentalcry.org/the-yurok-nation-just-established-the-rights-of-the-klamath-river/

34 Lake Erie Bill of Rights. https://beyondpesticides.org/assets/media/documents/LakeErieBillofRights.pdf

35 Brown, H. Claire. "How Ohio's Chamber of Commerce Killed an Anti-Pollution Bill of Rights." *The Intercept* August 29 2019. https://theintercept.com/2019/08/29/lake-erie-bill-of-rights-ohio/

36 Nathan Johnson, "SUVs are back, and they're spewing a boggling amount of carbon," *Grist* (October 18, 2019). https://grist.org/article/suvs-are-back-and-theyre-spewing-a-boggling-amount-of-carbon/

37 Amitav Ghosh, *The Great Derangement: Climate Change and the Unthinkable* (Chicago: University of Chicago Press, 2016), 1-84, 136.

38 Uri Gordon, "Darkness Falls: Revisiting anarchist politics in the age of collapse," *Medium* (August 9, 2019). https://medium.com/@uri.gordon/darkness-falls-revisiting-anarchist-politics-in-the-age-of-collapse-ac334d5f003e?fbclid=IwAR1j6hMm7f2hsUad1M-mP3aoXGcGdS758Y8SsDmSNGbWXTr9RPgPMI0Wcj0w

39 Simon Springer, "Anarchist Geography," *The Wiley-AAG International Encyclopedia of Geography: People, the Earth, Environment, and Technology* (2015), 4. https://www.academia.edu/6580928/Anarchist_geography

40 Kathryn Yusoff and Jennifer Gabrys, "Climate change and the Imagination," *Wiley Interdisciplinary Reviews: Climate Change* 2, 4 (July, 2011), 516–534, 524.

41 Morton, *Humankind*, 186.

42 Ibid., 173.

43 Ibid., 189.

44 Elise Amel, Christie Manning, Britain Scott, Susan Koger, "Beyond the roots of human inaction: Fostering collective effort toward ecosystem conservation," *Science* 356 (April 21, 2017), 275-179, 175. https://science.sciencemag.org/content/356/6335/275.abstract

45 Ibid., 276.

46 Ibid.

47 Ibid.

48 Ibid., 277.

49 Stevphen Shukaitis, *Imaginal Machines: Autonomy & Self-Organization in the Revolutions of Everyday Life* (London/New York/Port Watson: Minor Compositions, 2009), 104.

50 Brian Tokar, "The liberatory potential of local action," *Roar* (September 19, 2019). https://roarmag.org/essays/liberatory-potential-local-action-tokar/

51 Patrick Hayden, "Gilles Deleuze and Naturalism: A Convergence with Ecological Theory and Politics, 23-45: 35; In Bernd Herzogenrath, Ed., *An [Un]likely Alliance: Thinking Environment[s] with Deleuze/Guattari* (Newcastle upon Tyne, Cambridge Scholars Publishing, 2008).
52 Alf Hornborg, "How localization can solve climate change," *BBC Future* (September 6, 2019). https://www.bbc.com/future/article/20190905-how-localisation-can-solve-climate-change
53 Voltairine de Cleyre, "Direct Action." *The Anarchist Library*. https://theanarchistlibrary.org/library/voltairine-de-cleyre-direct-action
54 Murray Bookchin, *The Ecology of Freedom: The Emergence and Dissolution of Hierarchy* (Oakland, CA: AK Press, 2005), 48
55 Infoshop, "Direct Action." http://www.infoshop.org/direct-action/
56 Rob Sparrow, "Anarchist Politics & Direct Action," *The Anarchist Library*. https://theanarchistlibrary.org/library/rob-sparrow-anarchist-politics-direct-action#toc4
57 Ibid.
58 David Wallace-Wells, *The Uninhabitable Earth: Life after Warming* (New York: Tim Duggan Books, 2019), 13.
59 Peter Sloterdijk, *Spheres, Vol 2: Globes, Macrospherology.* Trans. Wieland Hoban (South Pasadena, CA: Semiotext (E), 2014), 940.
60 Ibid., 938.
61 Ibid., 951.
62 Michael Truscello, "The New Topographics, Dark Ecology, and the Energy Infrastructure of Nations," *Imaginations* 3, 2 (2012): 201. https://journals.library.ualberta.ca/imaginations/index.php/imaginations/article/view/27253
63 Mulford Q. Sibley, *Nature and Civilization: Some Implications for Politics* (Itasca, IL: F.E. Peacock Publishers, Inc. 1977), 229.
64 Springer, "Anarchist Geography," 4.
65 Daniel C. Wahl, *Designing Regenerative Cultures* (Axminster, England: Triarchy Press, 2016), 45.
66 Kieran Cashell, "Scared Shitless: Anarchism, Capitalism or Democracy," In Victora Brunetta and Kate O'Shea, Eds., *Durty Words: A space for dialogue, solidarity, resistance and creation* (Durty Books: Limerick Ireland, 2019), 112-113.
67 Kenneth M. Heilman and Russel S. Donda, "Neuroscience and Fundamentalism," *Tikkun* (July 2009). http://www.tikkun.org/magazine/tik0709/frontpage/neurosciencehttps://medium.com/insurge-intelligence/sustainability-is-not-enough-we-need-regenerative-cultures-4abb3c78e68b
68 Elizam Escobar, "Art of Liberation: A Vision of Freedom," In *Art on the Line, Essays by Artists about the Point Where Their Art & Activism Intersect*, Ed. Jack Hirschman (Williamantic, CN: Curbstone Press, 2002), 248-249.
69 Richard Gilman-Opalsky and Stevphen Shukaitis, 11.

Reworlding

1. Jack Thomas, "An apology from an environmentalist," *The Spinoff* (July 1, 2019). https://thespinoff.co.nz/science/01-07-2019/an-apology-from-an-environmentalist/?fbclid=IwAR388VQ6ei1KtbWdeYWnXAqPtykult6Alr4oL8Y5pwDi-D17BH1f7GzSWHA
2. Winona LaDuke, Facebook comment, December 19, 2019.
3. David Wallace-Wells, *The Uninhabitable Earth: Life after Warming* New York: Tim Duggan Books, 2019, 226.
4. Timothy Morton, *Being Ecological* (Cambridge MA: The MIT Press, 2018), 128.
5. David Wagoner, "Lost," In *Traveling Light: Collected and New Poems* (Champaign, IL: University of Illinois Press), 1999.
6. Daniel C. Wahl, *Designing Regenerative Cultures* (Axminster, England: Triarchy Press, 2016), 101.
7. Thom van Dooren, *The Wake of Crows: Living and Dying in Shared Worlds* (New York: Columbia University Press, 2019), 209.
8. Ibid.
9. Barry Lopez, *Of Wolves and Men* (New York: Scribners, 1978, 199. .
10. Personal communication, January 9, 2020.
11. James Koehnline, Facebook post, January 12, 2020.

Lightning Source UK Ltd.
Milton Keynes UK
UKHW022213220621
385988UK00005B/26